Death and Life of Nature in Asian Cities

Death and Life of Nature in Asian Cities

Edited by
Anne Rademacher and K. Sivaramakrishnan

**Hong Kong Institute for the
Humanities and Social Sciences**
(Incorporating the Centre of Asian Studies)

This publication has been generously supported by the Hong Kong Institute for the Humanities and Social Sciences, and it grows out of two intensive workshops organized by the Environmental Sustainability, Political Ecology and Civil Society research group in the Institute's Inter-Asia Program.

Hong Kong University Press
The University of Hong Kong
Pokfulam Road
Hong Kong
https://hkupress.hku.hk

© 2021 Hong Kong University Press

ISBN 978-988-8528-68-4 (*Hardback*)

All rights reserved. No portion of this publication may be reproduced or transmitted in any form or by any means, electronic or mechanical, including photocopying, recording, or any information storage or retrieval system, without prior permission in writing from the publisher.

British Library Cataloguing-in-Publication Data
A catalogue record for this book is available from the British Library.

10 9 8 7 6 5 4 3 2 1

Printed and bound by Paramount Printing Co., Ltd. in Hong Kong, China

Contents

List of Figures and Table	vi
Preface and Acknowledgments	vii
Introduction: Urban Nature Brought to Life in an Age of Loss *Anne Rademacher and K. Sivaramakrishnan*	1
1. Divine Excess: The Power of Accidents and Nature Spirits in Bangkok *Andrew Alan Johnson*	26
2. Putting the Garden Back: Cultivating Life through Urban Gardening in India *Camille Frazier*	44
3. The Village at the End of the World: Ecologies of Urbanism in Climate Crisis Imaginaries *Kasia Paprocki*	64
4. The Singapore "Garden City": The Death and Life of Nature in an Asian City *Annu Jalais*	82
5. The Absent Presence: Potholes in Urban India *Harris Solomon*	102
6. The Death and Life of Urban Ecological Commons in Taipei *Tomonori Sugimoto*	120
7. Making Land Out of Water: Ecologies of Urbanism, Property, and Loss *Shubhra Gururani*	138
8. Concrete Ecology: Covering and Discovering Saigon's Ecology in a Time of Floods *Erik Harms*	159
9. Keeping Pace with the Foodshed in Hangzhou *Caroline Merrifield*	177
List of Contributors	197
Index	199

Figures and Table

Figures

Figure 1.1: Dresses for the tree spirit	35
Figure 1.2: A dance as an offering	37
Figure 1.3: "Digging" for numbers	39
Figure 2.1: A rooftop terrace garden in Bengaluru	47
Figure 2.2: An Oota From Your Thota event	50
Figure 2.3: Srinath's "urban jungle"	55
Figure 2.4: Rajinappa's garden, located next to a major highway overpass	60
Figure 5.1: Pothole "puja" (worship)	111
Figure 5.2: Image of artist Rupali Gupte's installation "Pothole City"	112
Figure 6.1: Gardens on a hillside in New Taipei City	121
Figure 6.2: A riverside park in New Taipei City	129
Figure 6.3: Sra's rattan garden	134
Figure 7.1: Squatting and recycling in Ghata Lake–bed by migrant workers	152
Figure 7.2: View of the Grand Arch from Ghata Village	154
Figure 7.3: Facing Grand Arch	155
Figure 8.1: Pumping sand into wetlands in order to build the Thủ Thiêm New Urban Zone	161

Table

Table 7.1: Calculating water bodies	145

Preface and Acknowledgments

This book builds on more than a decade of collaborative efforts to better understand the interconnection of urban and environmental change. Since 2009, under the organizing conceptual frame, *ecologies of urbanism*, scholars have joined us to think more clearly about urban ecology, ideas of nature in cities, and questions of environmental sustainability in Asian cities. Across the pages of two previous volumes, *Places of Nature in Ecologies of Urbanism* (2017) and *Ecologies of Urbanism in India* (2013), we developed and refined a rubric for examining social experiences and enactments of nature within city spaces, and beyond them. We now add the present book and the rich, new insights of its contributing authors.

Through the ecologies of urbanism frame, we ask: how do we understand social change more fully, and even quite differently, when we consider human and nonhuman nature in an ecologically integrated analytical rubric? Our previously published volumes underline the ways that specific places and contexts matter, but they also emphasized consequential points of connection. These can include circulations of capital, labor, and information; they are also always nested into ecosystem scale biophysical processes. Our ongoing challenge is to see those consequential connections more clearly and to better understand how they drive social and environmental change.

This third project in the ecologies of urbanism effort traces its origins to a round table discussion at the 2017 annual meetings of the Association for Asian Studies. Our aim was to focus on urban decay, regeneration, growth, and sustainability as they arise in contemporary Asian cities. We noted that conventional urban studies research often narrates urban growth as a story of planning and disruption, the displacement of nonhuman life, or the depletion of human and natural resources to concentrate industry, commerce, and government.

We sought to explore a different analytical view, one that would consider the socionatural, historical, and cultural logics that underpin assessments of the death and life of nature in Asian cities. Hillary Cunningham, Will Glover, and Christina Schwenkel joined us to offer inspiring presentations. A lively discussion ensued before a large and engaged audience.

That round table, and the scholarly conversation it generated, led to a fully developed international scholarly workshop called, "Death and Life of Nature in Asian Cities." Through the support of the Hong Kong Institute for the Humanities and Social Sciences (HKIHSS), we gathered ten international scholars from a variety of disciplinary backgrounds to explore the ways that environmental dimensions of cities and urban life might also be understood in terms of death and life. We held that workshop in June 2018 at Hong Kong University, again through the generous support of the HKIHSS.

The scholars who gathered for that first workshop brought fresh questions, compelling case studies, and new insights to our organizing theme—the death and life of nature in Asian cities—but also to the ecologies of urbanism rubric itself. Energized by the momentum it generated, we then began working with Hong Kong University Press to bring this book to life. The contributions to this volume, and the conversations that contributors undertake across its pages and chapters, benefitted from a second, synthesizing workshop in September 2019.

As was the case in our first workshop in Hong Kong, our colleagues and graduate students also joined in our discussions to infuse new perspectives into the papers and to bring out their innovative contributions.

This book, then, is built on a firm foundation of collaborative support, contributions, and intellectual energy. It is the product of many minds and many hands, and those who have made rich contributions to its pages are many. We are mindful of their presence and are grateful for their gifts of time, patience, thought, and resources.

From the very beginning of our work to understand ecologies of urbanism, we have been blessed by unwavering institutional and financial support from the Hong Kong Institute for the Humanities and Social Sciences. We are deeply grateful to the director of the Institute, Angela Leung, for her vision, insights, and confidence in our *Death and Life of Nature* undertaking. The institute's former director and now honorary professor, Helen Siu, continued to provide essential moral and intellectual support to this third adventure on a rich intellectual journey, and her guidance continues to challenge and enrich our work.

The inspiring scholarship of HKIHSS-affiliated faculty and students brought many wonderful dimensions to our first workshop, and we are grateful to all who participated in those memorable days of paper deliberations and experiential field trips. We wish to thank Yvonne Chan and Joan Cheng for their impeccable administrative support: through their assistance, the Hong Kong workshop was a marvelous success.

Dorothy Tang and Amy Zhang offered excellent contributions to the proceedings of that workshop, and we are grateful for their participation and insights. Other preoccupations prevented them from joining the 2019 workshop at Yale University, and hence their contributions are not part of the volume. Their ideas and energy, however, contributed vitally to the fine start of this project in Hong Kong, and we remembered their papers with admiration as we moved forward to complete the book.

Together with the Hong Kong Institute for the Humanities and Social Sciences, Yale University has also long supported our ecologies of urbanism efforts. Through the Yale Inter-Asia Initiative, and with support from the Edward J. and Dorothy Clarke Kempf Memorial Fund, our second workshop was as successful as the first. Additional support was provided by the Council on East Asian Studies, the Council on Southeast Asia Studies, the South Asian Studies Council, and the Whitney and Betty MacMillan Center for International and Area Studies at Yale. The Department of Anthropology at Yale generously welcomed us for two fruitful days of deliberations. We remain appreciative of Marleen Cullen, Francesco D'Aria, and Jennifer DeChello for their logistical support.

We are also extremely grateful to Yukiko Tonoike, who through the Yale Inter-Asia Initiative has offered the kind of administrative, academic, and editorial support that has not only strengthened this book but that has helped us to forge the strong and longstanding bonds of a generative intellectual community.

The Yale workshop brought all the chapters into closer conversation with each other and helped to fine-tune our overarching arguments and framework. Our deliberations were enriched by the comments and questions raised by invited graduate students and scholars who took the trouble to read closely and offer valuable perspectives from beyond our author group. We thank Ritwick Ghosh (New York University), Adeem Suhail (Yale South Asian Studies), and several Yale anthropology students, including Bhoomika Joshi, Lav Kanoi, and Shoko Yamada, for their participation. An earlier version of the introduction to the volume also received useful comments from Dr. Sahana Ghosh and Dr. Sayd Randle, and we remain grateful for their engagement, which has made the Introduction stronger and crisper.

Finally, we are grateful for the ongoing interest and support of Hong Kong University Press and for the encouragement of editors Eric S. Mok and Kenneth Yung. Anonymous reviewers offered the kind of enthusiasm and constructive critique that energized and encouraged us, and we are now delighted to share this book with readers.

Introduction: Urban Nature Brought to Life in an Age of Loss

Anne Rademacher and K. Sivaramakrishnan

Introduction

The chapters in this book bring questions about ecological vitality into conversation with the everyday human experiences that shape the form and meaning of urban nature. Such a conversation reaches beyond the question of what "counts" as urban nature in a given time and place and invites us to think through the many nested scales at which nonhuman life, death, thriving, and withering come into being. Although the juxtaposition of death and life might suggest stark contrasts between beginnings and endings, our core challenge is to regard the making and unmaking of urban nature as a dynamic, ever-continuing, social-biophysical process.

Our collective starting point is the uncertainty, and yet generativity, of the Anthropocene—an era distinguished as it is by wholly new forms of nature, and long legacies of human nature making and re-making.[1] For some, it is a reminder of a more cataclysmic and epochal death of nature that might well be caused by hegemonic industrial-urban processes around the world (see Zalasiewicz et al. 2008; Bonneuil and Fressoz 2016; Malm and Hornborg 2014). Recognizing that nearly two decades of vigorous debate have scrutinized its underlying assumptions, we echo recent work by Dipesh Chakrabarty to note that conflicts between historical and geologic understandings of time remain at the heart of discussions and experiences of the Anthropocene (Chakrabarty 2018).[2]

Our engagement with this concept is informed as well by the work of Kathleen Morrison, who has noted its hidden Eurocentrism. Starting with European history as its prime temporal frame, she argues, has united the scientific community with humanists and brought otherwise disparate modes of thought into a unified attempt to fit social relations into determinative models. Such easy alliance across disciplines

---

1. See Waters, Zalasiewicz, Summerhayes, and Banrnovsky (2016); for a sense of what may remain possible on a planet fundamentally depleted by industrial civilization and its insatiable appetites, see Tsing, Swanson, Gan, and Bubandt (2017).
2. Earlier reflections on concepts and historical scope include Steffen, Grinevald, Crutzen, and McNeill (2011).

is harder to sustain when experiences and understandings outside of European trajectories are taken more seriously (see Morrison 2015).

In addition to historical and philosophical debates, as one of us has previously written, "there is also increasing disquiet in some quarters with the fact that some humans (very few as a proportion of the world population) created the problems, and the ones least responsible face the worst effects. . . . If humans are seen as one race, or species, brutalizing the earth, such a view obscures the reality of huge discrepancies that have emerged amongst humans in power and wealth over modern historical time" (Sivaramakrishnan 2019, ix–x). Arguably, these profound inequalities, amplified by the death of nature writ large, are strikingly evident in urban life. This is vividly so in Asia.[3]

In invoking the death and life of nature in cities, then, we remain mindful of the ways that the very concept of the Anthropocene hinders a discussion of environmental justice issues. This becomes even more salient when our notion of justice extends to the rights of nature, and to analyses that focus on the varieties of nonhuman lives threatened or expelled in the making of modern cities and towns. This book is thus also a call to reflect upon, and to anticipate, how we might more carefully study nature and contemporary city life as continuously co-produced. This requires, we argue, keeping inter-scalar relations, spatial variations, and social differences firmly in our sight to discern where and how shared ecological and social processes generate distinct experiences.

Examples within twenty-first-century Asian cities include the uneven distribution of health risks posed by air pollution, struggles for potable water, and large quantities of urban waste in cities like Delhi or Beijing. While waste, for example, may generate precarious livelihoods, it can also present immense hazards to the poorest workers who handle its disposal or live in its close proximity. In the case of Phnom Penh, we might note that even a well-articulated sewage treatment system produces dangerous marshlands, offensive odors, and organic material in which poor squatters grow a variety of vegetables (Jensen 2017, 636–37). At least in South Asian cities, emaciated cows and goats might be seen jostling for space with well-fed and beautifully groomed thoroughbred pets. Uneven distributions of, and access to, green spaces in cities across Asia serve as reminders of a twofold exclusion: the social exclusion of underprivileged city-dwellers and the subordination of plants and fauna to the crafted aesthetic of the verdant Asian city.

The Death of Nature Re-imagined, in the City

In many North American academic discussions in the field of environmental studies, the phrase "death of nature" may call up visions of decades-old warnings,

---

3. For a sweeping and compelling account of growing economic disparities and their effects in the world, see Piketty (2014).

once traced by the likes of Rachel Carson's *Silent Spring* (1962), Bill McKibben's *The End of Nature* (2006), or Carolyn Merchant's *The Death of Nature* (1990). More contemporary students in that same field may think first of public intellectual voices like those of Elizabeth Kolbert (2015) or Amitav Ghosh (2017). While the earlier warnings sounded by writers like Carson were, in many ways, genesis narratives for entirely new political missions, social collectives, and, indeed, academic subfields, we notice that "nature," in that earlier historical moment, was still largely regarded as the thing that stood in opposition to the city. In dominant Western sociocultural imaginaries, nature in that era was not only absent from cities but completely separate from modern social life. Nowhere was the death of nature more visible, in this mode of thinking, than in cities.

Today, across Western academic and popular discourses, the conceptual content of "nature," and our ability to trace its existence in social life even in the most densely populated cities of the world, has changed profoundly. We now hesitate to separate human beings from nature, ecology, or the environment, and in fact recognize "urban ecology" as an arena for substantive inquiry rather than a puzzling oxymoron. And yet, we still inhabit a planet whose ecological present is perhaps most accurately conveyed through human accounts and experiences of loss and death—whether they take the form of our expectations of eco-collapse and catastrophe, or they fuel urgent missions to shore up something called "resilience."

In this *Sixth Extinction* era marked by what Ghosh (2017) has characterized as a "great derangement," we are acutely aware of the biophysical costs of our long global history of empire and industrialization. New accountings of loss emerge at an almost dizzying pace, documenting mass die-offs,[4] oxygen-starved oceans, entire islands of plastic waste, and an irreversible loss of unfathomable biodiversity.[5] At this writing, cities worldwide are held in the grip of a global pandemic that has brought renewed attention to our interactions with the nonhuman world, and how dramatically changed ecologies and ecosystems can always introduce new consequences—some involving previously incomprehensible human social experiences of loss.

We begin our exploration of *Death and Life of Nature in Asian Cities*, then, by noting that in scientific, social scientific, and cultural registers, the death of nature, however we may have reconceptualized nature itself, is perhaps experienced now with more gravity and socio-ecological traction than ever. We further suggest, in the spirit of Bruno Latour's now-classic investigation of the modern historical human subject, that our studies inevitably confront the predicament of the Modern (in his use of the term), whereby those who have spread out to fill all available space now find themselves lacking room. In his account, this quest for space, which he also renders as a place to situate existence in the world, leads to a "groping in the

---

4. For example, Hallman et al. (2017).
5. Kolbert (2015) estimates flora and fauna loss by the end of our century to be between 20 and 50 percent of all living species on earth.

dark," for it reveals the sense of unsettled feeling that comes with "brutal expulsion . . . from the entire habitable Earth" (Latour 2013, 104). For us, Latour is not only advocating an anthropology of the Anthropocene era's predicament of modern humans, but he is also leading us toward a mode of inquiry that is deeply informed by the emergent concerns of the environmental humanities.

What do we mean by the environmental humanities? In the inaugural issue of the journal *Environmental Humanities*, Deborah Bird Rose and her colleagues offered a useful set of programmatic definitions that guide us. They note how much environmental research, even in the human sciences, previously proceeded with a "narrow conceptualization of human agency, social and cultural formation, social change, and the entangled relations between human and nonhuman worlds." They suggest that the development of the environmental humanities might "enrich environmental research with a more extensive conceptual vocabulary, whilst at the same time vitalizing the humanities by rethinking the ontological exceptionality of the human." Specifically, "the humanities have traditionally worked with questions of meaning, value, ethics, justice and the politics of knowledge production. In bringing these questions into environmental domains, we are able to articulate a 'thicker' notion of humanity" (Rose et al. 2012, 1).

Apart from scientific and policy responses to global environmental change, environmental humanities responses tend to focus on fundamental questions of meaning, value, responsibility, and purpose. Since the early 2000s, the physical science account of the Anthropocene is one in which the earth and its atmosphere have been irreversibly altered by human influences on biogeochemical processes.[6] For the humanities, as one set of scholars of literature put it, this has meant that the scholarly imagination now operates on "a wholly different scale, vastly more global in scope, vastly more historical in extent . . . and . . . that [takes] seriously the specific responsibilities that arise from this shifting of perspectives" (Garrard, Handwerk, and Wilke 2014, 149).

Yet we note that capacious notions of humanity that never excluded the nonhuman persisted across Asian societies and continued well into the onset of industrial modernity.[7] In the study of South Asia, the region with which we are most familiar, we would suggest this with even greater assurance, as in that region a thicker notion of humanity was never entirely removed from scholarly perspectives and analytical procedures. Subjects like the deep history of civilizational processes, the colonial encounter, pantheistic or world-renouncing religion, human relations with animals, and the picturesque in the history of art afforded opportunities for exploring the environmental humanities without giving it that name.[8]

---

6. For some of the earliest and authoritative formulation of these ideas, see Crutzen and Stoermer (2000); Crutzen (2002).
7. Examples of this argument can be found in Duara (2014), Elverskog (2020), and Sivaramakrishnan (2015).
8. Recent path-breaking work in art history, for instance, has brought human/nonhuman relations in India into environmental humanities at the intersection of religious studies and studies of architectural history. See Ray (2019).

For instance, a robust literature on environmental history emerged in South Asian Studies by the 1990s. This literature often converged with the writings of early environmental activists like Anil Agarwal to argue that nature conservation must move beyond "pretty trees and tigers" to questions of social equity. Destruction of habitats, or biodiversity, also placed the lives of millions of *Adivasis* and other politically or economically vulnerable South Asians at risk. In one sympathetic line of thought, this mode of analysis also enabled the "discovery" of the environmentalism of the poor (see Guha and Martínez-Alier 1997; Martínez-Alier 2002; Peet and Watts 2004). We understand the environmental humanities, then, to denote a moral, political, and artistic commitment that is extraordinary and generative.

Our exploration in *Death and Life of Nature in Asian Cities* aims to take seriously the idea that "a reflexive, anthropocenic cultural politics" (Garrard, Handwerk, and Wilke 2014, 149) can potentially get us beyond the mess created by an anthropocentric hubris that informed the conquest of nature in modern times, especially in the twentieth century. This is also a potential platform on which to unite across the disciplines and epistemic divides that separate scholarly inquiry.

It is in this spirit of inquiry that this book considers the death and life of nature in Asian cities. Its chapters present an array of contexts within which human communities experience the complex, changing webs of urban life. We fix our social and spatial focus on cities: after all, "nature" is not the only concept that has undergone significant rethinking over the past decades. Cities, too, no longer sit comfortably removed from their hinterlands; we now understand them to be integrally connected to far more complex tendrils of exchange, movement, and flow. In their reconceptualized guise, cities are more often understood as nodes along a far more expansive urban continuum—a continuum that encompasses the full range of landscape densities, social communities, and biophysical systems.

If we follow Lefebvre ([1970] 2003) to regard the world as now "completely urban," that is, interwoven in varying ways within threads of late capitalism, then studying nature in cities challenges us to trace their contours across the spatial and social processes that span this continuum; these often include interconnected hinterlands with as much complexity as they do cities themselves. This is as true in social and material terms as it is in biophysical ones: the biogeochemical cycles that undergird any city's ecology rarely start and end with municipal boundaries.[9]

We are also mindful that in the present, neither city infrastructures nor their human populations are necessarily fixed—in space or in place. As we witness the global movement of those displaced as environmental refugees, trace the migrations of those fleeing overt violence and war, and note the complex and varied dynamics of displacement and migration, we are roundly challenged to encounter a planet of cities that is not only rapidly urbanizing but very much "on the move"—marked by

---

9. Fine work on ecological footprints of cities, even premodern urban concentrations, has begun to take scholarship to this realization based mostly on work in Europe. See, for instance, Hoffmann (2007).

large movements of migrating peoples in the present, and likely to remain so with evermore intensity.

Cities are also potentially on the move in material terms: regardless of their location, all cities face a present and future scenario punctuated by potentially profound ecological transformation as climate change introduces new stresses and intensifies existing ones. Here, we are reminded of a very recent report by the Intergovernmental Panel on Climate Change on sea level rise (see Oppenheimer et al. 2019). The overwhelming prevalence of large, dynamic, populated cities in coastal areas and the remaking of water-land boundaries that sea level rise ensures come together in this study to underscore the dynamism and uncertainty that is present in the material, built infrastructures that constitute a vast majority of city territories. We therefore must anticipate a coastal city map that is nothing if not dynamic, while we also attend to the uneven human capacity to respond. To consider the death and life of nature, then, is also an invitation to seriously consider human social experiences of vulnerability, risk, and marginality, and to see in new ways how everyday ideologies of social belonging carry profound social and biophysical consequences.

It is in these dynamic circumstances—the lived, everyday life of cities in our time—that we have assembled this volume of careful, case-based work. Our aim is to better understand the coproduction of social and biophysical nature (see Rademacher, Cadenasso, and Pickett 2019), and, we hope, the dynamics that join ecological vitality and human well-being together.

As has been the case across our two previous *Ecologies of Urbanism* volumes, we focus here specifically on Asia as a region in which cities often invoke a palpable sense of becoming. From Asian hyper-cities to smaller cities and towns, these are the urban landscapes that host fourteen of the world's twenty-eight megacities (ten million or more) and that, by 2030, are expected to host twenty-four. When combined with 330 medium sized cities (1–10 million) and 815 smaller cities (with populations of 300,000 to 1 million), Asia emerges as the global center of city-making.

Not unlike the cities of Europe and North America that took shape over the nineteenth and early twentieth centuries, Asian cities emerged in their first modern avatar during the twentieth century: they embodied industrial ambition and governmental power. In so doing, they commissioned the death of nature in many ways. In many situations industrial cities contaminated and killed nature in the form of land and water, but these essential elements returned in their poisoned state to endanger human life and render it painful and insecure. One powerful example of such processes—encompassing mining waste, factory effluents in rivers, and air pollution in studies of Asian urban industrialization—is that provided by Brett Walker (2011) for Japan.

For another example, consider the industrial towns and cities along the middle Ganges in the north Indian plain. These discharge waste, sewage, and factory effluents that often render river water dangerous even for irrigation, let alone bathing or drinking. This condition persists despite four decades of projects that were designed and executed to control and diminish the pollution of the national river, which is also a sacred river for India's majority Hindu population (Sen 2019, 351). As a recent study of waste and garbage effectively shows, urban refuse in places like India has grown exponentially even as its inorganic composition has rapidly increased. Meanwhile, older systems of collection and recycling have collapsed under the weight of consumer culture, municipal incapacities, and toxic detritus that suffocates many of the channels that nature provided for digesting and removing waste (Doron and Jeffrey 2018).

City infrastructural projects have long sought to contain rivers, build waterworks, and denature the biological life of water bodies in the name of purification, or by treating them as waste disposal sites. Such was the plight of the Thames river in London, for instance, in the aftermath of the Second World War. In 1959, the *Manchester Guardian* called the Thames a badly managed open sewer; its biological oxygen demand under London Bridge was virtually nonexistent (Hardach 2015; see also Kelly 2018). It took another twenty years, and a vigorous effort by government and environmentalists, to clean the river. The Thames now teems with aquatic life and fresh water, as well as opportunities for London residents to find respite or meaning in the urban nature of the river that is inseparable from their city. Many other European rivers can recount similar histories of the death and life of nature.[10] But far fewer such stories yet exist for Asian cities.

As cities in Asia grow and proliferate, infrastructural systems pipe water from rivers and ground sources, suffused with chemical cleaning agents, into homes and other buildings. Along the way, marshes and wetlands are often drained or covered in order to build factories, dwellings, harbors, and commercial hubs. In her study of city-making in colonial Calcutta, Debjani Bhattacharyya describes how such marshes and deltas carried the imprint of premodern engineering, while modern urbanism attempted a more radical transformation of land-water relations. This created an unprecedented killing field for nature by the middle of the twentieth century (Bhattacharyya 2018).

In the same way, colonial and modern city-making in Asia affected soil and vegetation complexes, often the habitat of microfauna and host to rich biodiversity. These were subordinated or submerged below tar and gravel to create roads and highways. Animals that signaled the presence of wilderness or agrarian life in spaces designated for cities were either expelled or displaced to the terrain outside of municipal limits. It is these kinds of enclosures, expulsions, and zoning that demarcated and built a sanitized, human-designed, and human-dominated norm for modern city life in Asia.

This pattern of modern Asian urbanism seemed to emulate at a more breathless pace what had already happened in industrializing urban centers and metropolitan

---

10. For a fine study in this vein, see Cioc (2006).

hubs of modern Europe and North America. In the cities that emerged across Europe and North America by the twentieth century, domestic and work animals found themselves increasingly unwelcome. Thus, when it became hard to conceive a safe city for horse and human, the quadruped was displaced to farm suburbs (see McShane and Tarr 2007). In a classic study of Seattle, we learn that various animals were essential to the building of the modern American city in the late nineteenth century, and the removal of such animals was linked to the formation of new middle-class neighborhoods during the twentieth century. As older relations with livestock became more remote, domestic pets emerged as even more important human companions, perhaps another mark of the many deaths for nature, and its many forms, in the histories of America's modern cities (Brown 2016).

Arguably, even as certain animals have returned to cities—by invitation or as opportunistic but unwelcome co-residents—their place in a culturally sanctioned pantheon of living nature in a city environment remains a topic of intense contestation in the USA. We have only to consider the complex urban regulation of farmyard animals as distinct from pets, animals raised for sport and recreation, companion species, and the proliferation of organizations that advocate for and against their fair treatment or rightful place in the city in order to appreciate this point (see, for instance, Haraway 2003; Mullin and Cassidy 2007). This is true in the cities of Asia as well, but the temporalities of animal exclusion differ considerably across cities in North America, Europe, and Asia.

In much of the European and American experience, animal displacement often preceded growing social awareness of industrial pollution. Such animal exclusion has been at the heart of the remaking of nature in the city. One broad trend described by historians of major cities like Chicago, New York, and Seattle is the remaking of livestock-friendly towns of the nineteenth century into pet-friendly but livestock-averse cities by the end of the twentieth century (Brown 2016; Van Horn and Aftandilian 2015).

In Asian cities, these issues often intertwine as part of a late twentieth-century "cleanup" of cities energized by urban environmental movements, governments, and vocal middle-class activists. For example, the plight of cows in Delhi is instructive: having lived in the urban villages that were enclosed as Delhi grew, cows and their *Gujjar* or *Ahir* keepers were key to the city's milk supply into the 1970s. Then the government of India invested in a massive program for the development of national milk production and distribution, which came to be known as Operation Flood. In cities like Delhi, booths vending milk sprung up under the brand name of Mother Dairy. As this system grew into the familiar neighborhood source of safe, refrigerated milk, local *gwalas*, with their wandering milch cattle, were rendered a nuisance—in other words, unwanted nature.[11]

---

11. For more on the campaign against stray cows in Delhi, see Baviskar (2015).

In the late 1970s cleanup, resettlement colonies landed in the outskirts of Delhi as part of slum clearance.[12] By the 1990s, the unbridled growth of the city—west, north, and south (the river bounded it to the east)—took it further into the semi-arid scrublands of Haryana and the fertile farms of western Uttar Pradesh. There, monkeys and displaced poor Delhi slum dwellers were forced out together. For the monkeys, removal was a sanctuary they resisted: they returned to the city at will (Gandhi 2012, 47–48). The urban poor found it much harder, as new and unprecedented numbers of migrant workers surged into Delhi city, finding work in construction, domestic service, and other service sectors.

Considered together, these examples from America in the nineteenth and early twentieth centuries and from India in the late twentieth century illustrate how nature was controlled, subjugated, disciplined, removed, or transformed into built city environments that seemed, through manifold deaths of nature, to garner life forces of their own. We might see these forces reflected in engineering marvels, planned vistas, and thoroughly socialized spaces in cities and towns. At the same time, our aggregate, totalizing account is also too simplistic. City-making endeavors were never complete or without contestation. Forms of nature—whether fugitively profuse, tidily manicured, or serviceable in some fashion—remained in these cities.

Their manifestations could be as clusters of unwanted plants, the presence of disease-carrying bacteria, compostable food waste, enduring enclaves of animal life, urban parks and protected areas, green belts, and even avian life adapted to ornamental or fruit trees along avenues and in-home gardens. And nature reappeared as water seeping, overflowing, stilled, in scattered wet zones or flood events, with co-resident fungi and moss, and in the organic matter, or solid waste, that sullied the quality of water in particular places. What we may consider the recalcitrant lives of nature, then, were always at risk, and fitfully being banished. Nevertheless, they persisted.

We've begun our exploration of the *Death and Life of Nature in Asian Cities*, then, by grounding our thinking in historical conditions. This points us toward the political work that was always necessary in order to enact ideal urban visions. It also highlights the subjectivities that individual agents have long brought to the praxis of urban transformation, and to the power relations that organize and legitimate environmental knowledge itself. Contemporary aspirations for urban sustainability and livability are always cast against this uneasy backdrop of historical forms of loss and instability.

What may arise today as excessive pollution and solid waste, water scarcity, insufficient housing conditions, crumbling or absent infrastructure, or—in the most contemporary sense—our realization of the profound vulnerability of coastal cities to sea level rise and catastrophic storms will always have their consequential historical contexts built into their possibilities for the future.

---

12. Tarlo (2003) provides a searing account of the human displacement initiated by slum clearance drives in the late 1970s.

In this book, our contributors do not go into the many epidemiological aspects that the death and life of nature in cities invites. It is nevertheless possible to ask how bursts of city development, with attendant piles of refuse, stagnant water, uncovered ditches, and putrefying solid waste create new hazards and patterns of withering; they nurture water-borne diseases, for example, that can rapidly spread in dense Asian cities (see Gandy 2014, 5–6; also, Dasgupta 2012). Many contemporary Asian cities are now embarked on ambitious projects of greening, climate change adaptation, resilience, and restoration precisely in order to fight human health threats like epidemic disease or deadly air pollution. To that extent they are also rediscovering nature in a salubrious, even life-giving, condition.[13]

Lively Urban Nature

Our previous volumes in the *Ecologies of Urbanism* series—*Places of Nature in Ecologies of Urbanism* (2017) and *Ecologies of Urbanism in India* (2013)—helped us organize a systematic exploration of the ways that historical and anthropological sensibilities can be fruitfully combined with engaged attention to the transformative power of nonhuman biophysical processes. In those books, contributors asked: how do we understand social change more fully, and even quite differently, when we consider human and nonhuman nature in an ecologically integrated analytical rubric? Both volumes underlined the ways that specific places and contexts matter, as do the connections forged through flows of capital, labor, and information. To these, we were careful to add the ecosystem-scale transformations that enable or disable those same flows.[14]

In this third *Ecologies of Urbanism* volume, our contributors begin with the assertion that fundamental questions of decay, regeneration, growth, and sustainability sharpen our understanding of social and ecological transformation. Our goal is to see more clearly how nature is made in simultaneous biophysical and cultural process, and how, in that interlinkage, multifaceted death and life are experienced in Asian cities. We assert that topics like sustainability and resilience, which in their most current manifestation are often expressed in ideas like green or smart cities,

---

13. Such projects of greening are driven by government programs and business initiatives, as well as growing demand from burgeoning urban middle classes across Asian cities. One recent examination of efforts largely in the corporate sector can be found in Clifford (2019).
14. The analytical rubric of *Ecologies of Urbanism* proceeds from the following assertions:
    - Analytically, "ecologies" is plural: there are many ways of knowing nature.
    - Biophysical conditions and their histories matter, including the colonial spatial and social legacies that are woven into urban landscapes.
    - To understand the city, one has to start with processes, not borders: an urban watershed, a network of tubewells, or an island that is the product of land reclamation are all crucial aspects of "the city" as an analytical category, not just as a territory on a map. This helps us to understand cities as more fluid than their conventional boundaries might suggest.
    - The ecologies of urbanism rubric would recognize a rail corridor as potentially fruitful a site for studying coupled biophysical and social transformation as the large "cities" that they connect as they traverse the landscape.

are about rediscovering human life itself as a multidimensional experience in which varied forms of nature are integral to reimagining urban existence.[15]

Sustainability plays a central role in these considerations. As Paul Warde reminds us, in English usage, the very term "sustainability" is remarkably recent, though its antecedent ideas and practices can be traced to early modern European history. It is, as he further observes, a profound notion that identifies ecological foundations of civilization that must be renewed, protected, and allowed to flourish (Warde 2011).

We do not need to dwell here on the truly vast literature on how the governance of nature became entangled with ideas of social improvement, technological advance, and greater efficiency in the use of natural resources—all classic processes bound up with the rise of modernity in Europe and Asia. It is important to note, however, as Warde does in his comprehensive study, that sustainability grows out of a command-and-control perspective and into a way of being and thinking in nature that recognizes its vastness, richness, and ability to confound human projects of the greatest sophistication (Warde 2018, 11 and 314–49). Such recalcitrance in nature may be feral, as some argue, or manifest in collapsing systems that leave humans vulnerable to previously unimagined danger.[16]

The agency of nature may be celebrated, and even romanticized, in some of these accounts, but we aim in this project to anchor our analyses to wider ecological relations and their uneven production of livelihood, amenities, rights, and justice in cities. This follows provocations from scholars like Amita Baviskar, whose critique of urban environmental activism notes that "ecology is often hard to see in the city. The concentration of concrete and tarmac, brick and glass, seems to squeeze it out of existence. The built environment overwhelmingly appears to be an artifact of human manufacture, of materials transformed by technology" (Baviskar 2018, 90). She is rightly urging us to look beyond ornamental or aesthetic values of nature in the city to its ecosystem functions, unequal accessibility across social classes, and privatization where it might otherwise be part of urban commons in ways that impinges on its ecological value to sustainable urbanism across social groups.

Our perspective on ecologies of urbanism also departs somewhat from more conventional urban ecology. Some work has followed one influential approach, developed in the eastern USA, which traces the shift from the study of ecology in cities to the ecology of cities.[17] In this approach, urban systems are viewed from an ecological perspective, thus including the natural and built environment and all resource use within the urban domain in an integrative framework. Ecological footprint analysis and industrial ecology are two ways in which such ecosystem

---

15. For a consideration of the idea of smart cities and its influence in urban planning and policy in India, see Sharma (2018).
16. For the idea of feral plants emerging in conditions of extreme ecological stress, see Tsing, Degeret al. (2020); and Tsing and Bubandt (2018).
17. This pioneering work and resulting framework for urban ecology studies is reported in Grove et al. (2015).

approaches have been fruitfully deployed in the study of cities and other urban concentrations. Considering cities as urban ecosystems, however, is often limited by collaborations that foreground the natural sciences, social sciences, and policy sciences (Grove et al. 2015, 2–14).

These approaches may sometimes conceive of ecology as a set of closed loops, and configure singular connections and flows. A singular ecology, in this kind of analysis, provides a neat system, but it also requires erasures and omissions. By extending our understanding of urban ecology to humanistic perspectives that encompass questions of affect, meaning made in lived experience, and the politics of representation, we employ a framework that pluralizes ecology and permits an examination of ecologies—understood as multiple webs of dynamic connection across human and nonhuman worlds. These webs may be emergent, contested, and often expressive of divergent patterns and processes of change, adaptation, transformation, and restabilization. These are also themes central to the environmental humanities perspectives, which we discussed above.

The death and life of nature, in the ways we have been describing it, may in fact have already captured the attention of some ecology advocates. For instance, if we consider two South Indian cities—Chennai and Bengaluru—we might point to examples in the form of tanks in one case and lakes in the other. Both tanks and lakes were part of each city's historical water supplies and built environments. However, in many cases, they suffered a precipitous decline after municipalities took them over. City construction, discharge of effluents, and waste dumping intensified, with tanks and lakes serving as receptacles for the refuse of city life or making way for urban infrastructure like roads, shopping complexes, and institutional facilities.

Urban water tanks assumed a prominent place in the oldest parts of Chennai. Most were likely to be part of a given neighborhood's temple complex. If the *agraharam* stretched in front of the entrance, the tank lay behind it. The Kapaleeswarar Temple, *agraharam*, and tank in Mylapuru is one emblematic case of this landscape design.[18] Such tanks were both a communal resource and a sacred space. The decline of the tanks, then, coincided with congestion and profanation. To the extent that they have been rejuvenated, it has been through social acts of reclamation and purification. The democratization of temple entry—as a consequence of temple entry movements in South India that secured access for lower castes over the course of the twentieth century—has played no small part in this revival. Temple revenues increased, they were renovated, and the tanks were revived.

Other factors include a surge in popular Hinduism and visits to temples, with growing interest (even from nonresident Indians) in the upkeep of temples in what they consider to be hometowns or villages (see the key work in this regard by Fuller 2003). The growing prosperity of temples in major cities is also facilitated by visits

---

18. The *agraharam* refers to the residential enclave of the Brahmin community in close proximity to the temple, often in the heart of the city, and in fact abutting the outer walls of the temple. Many of the temple priests officiating at the temple would live with their families in the *agraharam*.

from migrant and other professional transients in the city, and the willingness of temple communities to innovate rituals and organize festivals that allow a broader temple-allied public to form and contribute to upkeep.[19]

In the case of Bengaluru, the intertwined functional and moral values of urban water bodies are essential to ecological and cultural ideas about the overall vitality of the city. Thus, Harini Nagendra writes that, "within a few decades, the culture of valuing lakes as water-providing reservoirs, and reverence for lakes as life-giving, sacred entities, have given way to the use of lakes as septic tanks and garbage dumps" (Nagendra 2016, 173). Yet this death of a very specific form of nature has also sparked community-based efforts at resuscitation. In 2015, the organization *Jal Mitra* began to conserve a lake in North Bengaluru and now, six years later, it has grown to act for the revival of other water bodies in the city as well. The group's efforts encompass environmental education, corporate social responsibility projects, and urging the city to execute public works as nature-sustaining rather than nature-destroying projects (Gajjar 2016).

These very brief examples highlight some aspects of what may be considered "post-industrial ecologies of urbanism"—at least to an audience of scholars of the global north. After all, in American cities, the vitality of cities of the future often implies the return to life of certain forms of nature—particularly parks, open spaces, green streets, and community gardens. As Sarah Charlop-Powers (2016), executive director of the Natural Areas Conservancy, writes: "many have come to expect that our urban parks should provide residents with a broad suite of services above and beyond recreation—including flood protection, clean air and water, biodiversity, and respite from the pressures of urban life."

Such revisioning of urban nature as a source of ecosystem services and solace in the city is not absent in the South Asian experience, a topic addressed to some extent in work one of us has done on green spaces in Delhi (Sivaramakrishnan 2017). It can also be found in the planning and developing of "greener" cities across Asia. In this sense, the work in this volume emphasizes the need to consider the production and presence of lively urban nature as a core aspect of varied ecologies of urbanism.

We offer, therefore, an expansive definition of urban nature that is undoubtedly indebted to ideas about urban ecology, environmentalism, and cities as multispecies habitats. We wish to keep in focus the uneven distributions of amenities afforded by forms of city nature, and the ways that rights in, and to, the city are mediated by natural assets and their accessibility. Our concept of urban nature encompasses a material and imaginative realm of human and nonhuman sociality, comprising both mutualism and antagonism that is historically shaped and constantly reshaped in the city.[20]

---

19. An excellent account of such processes may be found in T. Srinivas (2018).
20. We are in agreement with Maria Kaika in many respects when she says urbanization strives to "render cities independent from nature's processes" but ends up tying nature and the city into a "socio-spatial continuuum"

The contributions to this book illuminate a continuing process of discovery and renewal through which urban natures may well be moving from taken-for-granted infrastructures to more consciously observed and experienced interplays between nonhuman existence and daily life experiences. These provoke moral and ethical evaluations of the human ecology of city life and direct relations between nature and culture into new avenues like aesthetics, care, perception, and stewardship. The book finds nature as often in a walk to the bus stop as in a stroll in a park.[21] It follows what Esther Woolfson (2014) has characterized as the urban natures that emerge through cultivated sensibilities. For some, these are available from unforgotten traditions, and for others they emerge from learning to become a natural historian and to be alert to nature in manifold forms and surroundings.

We find these conversations emergent among scholars of the environment from diverse backgrounds. The work in this book is conversant, for example, with a set of essays by Terrell Dixon. There, Dixon (2002) shows how rivers, parks, vacant lots, lakes, gardens, and zoos convey nature's rich disregard of city limits signs, each in their own different ways. New ideas about resilience and sustainability follow. Though Dixon's exploration of urban nature may draw from styles of documentation found in Europe or North America, they are not unfamiliar to Asian contexts.

Globally circulating modes of cosmopolitan urbanism related to environmentalism, conservation, organic farming, greening urban space, and so on are of course available to many residents of Asian cities and towns. Such practices also emerge, however, from distinctive modalities of Asian city life created in historically layered urban experience: these encompass empires, colonial encounters, and global city building across socialist, capitalist, or neoliberal regimes of investment and urbanization.[22]

The chapters to follow, then, trace some of the myriad ways that city dwellers formulate and experience instances of urban nature, its passing, and its renewal. They highlight material and imaginative processes that are unfolding in diverse patterns across Asia. One such process is the rapid urbanization of Asia across big cities, smaller towns, and the newest urban concentrations that enclose farms, woodlands, or other land uses.[23] The other is the contentious debates and novel schemes by which nature—all of nonhuman life and the inanimate world—is figured and emplaced in cities and their conurbations.

In the growth of cities and towns, and in their greening, we wish to underscore how new possibilities for foregrounding environmental or sustainability goals can

---

(see Kaika 2004, 5). She is, of course, working in a renowned Marxist tradition in thinking about urban nature. This might be traced to David Harvey (1996), who argued that cities are dense networks of interwoven sociospatial processes that are simultaneously human, material, natural, discursive, cultural, and organic.

21. We are inspired here by work like Johnson (2016).
22. One example from India would be Krishen (2006).
23. A good example of how urbanization with all these elements also produces vivid disparities in the way modern urbanism is experienced and described across social classes may be found in works like Harms (2016), Ghertner (2015), Searle (2016), and Schwenkel (2020).

emerge—be they in the form of green architecture as an emerging field of urban practice in Mumbai (see Rademacher 2017), or the greening effects of verticality in dense, undulating, and compressed land masses like Hong Kong (see Shelton, Karakiewicz, and Kvan 2010).

The chapters in this book also amplify how the intersection of urban growth and urban nature is a place rich with fresh ideas about urban planning, governance, and social life. It is a site of novel forms of politics and new subjectivities that individual agents bring to the praxis of urban transformation. Yet these are always in tension with new, and newly legitimated, forms of social exclusion and boundary-making. Moreover, when nature is valued in specific ways—for example, by an ecosystem service provider in a city aspiring to resilience or aesthetic appeal—we find that while nature itself can take on fresh meanings for urban residents and planners, so too can specific, often already-marginalized, populations.[24]

Each chapter of the book presents a case study, and together they survey cities across Asia. Drawing on careful historical and ethnographic approaches, each addresses relationships between the form of the nonhuman environment and the composition of social life. Each notes how certain aspects of urban nature are socially valued, while others may be deemed disposable. Our contributors ask, what socionatural, historical, and cultural logics underpin assessments of the forms of life that are permitted to live and thrive in the cities of the twenty-first century? The cases often show that human ideas and social agendas that galvanize around urban nature can mediate our understanding of precisely what, and who, is entitled to live and thrive in the city, and who or what is committed to social or species death.

It is our contention that across Asia, especially in crowded metropolitan centers, but also in smaller towns and at urban margins, the hazards, infirmities, and sources of vitality to be found in city landscapes are encountered and experienced differently across varied social groups. As cities respond to the degrading effects of urban development with ambitious greening projects, they may institute a second displacement or a double dispossession of the urban poor or migrants, whose livelihoods are often newly marginalized.

Nature in these settings returns in new ways to join assemblages of articulation in fresh struggles over rights to the city, even in forms we may not, at first, expect. For example, in the essay by Andrew Alan Johnson[25] in this volume, it is supernatural beings that haunt bridges, flyovers, and construction sites. These remind us that in alliance with those who suffer, submerged land and felled trees can author a moral critique of exclusive urban planning and prosperity. They may even do so in the form of ghosts, spirits, and locally powerful deities who need to be propitiated

---

24. A fine example of such a search for resilience, comparing an Asian city (Chennai) and a European metropolis (London), can be found in Niranjana Ramesh's unpublished PhD thesis (2018).
25. See Andrew Alan Johnson, "Divine Excess: The Power of Accidents and Nature Spirits in Bangkok," chapter 1 in this volume.

by those who would benefit from the infrastructures that have been built by means of their destruction or displacement.[26]

A contribution to our collection by Tomonori Sugimoto[27] explores how rights to the city may be juxtaposed against greening projects that define classed and privileged access to urban common spaces in Taipei. Here, certain practices—in this case, the provisioning of food and shelter along riverbanks by indigenous Taiwanese—are rendered increasingly invisible and undesirable. Like the marginalized populations in Bangkok about whom Michael Herzfeld has written, Sugimoto shows how notions of entitlement to territory involve a complex bundle of political rights, moral prescriptions, and notions of ritual belonging among native Taiwanese.[28] Urban nature and a transforming underlying ecology are lively collaborators here, as the biophysical landscape is changed and cycles of ritual and meaning-making transform. This chapter nests the cycles of social and natural meaning-making in the shifting landscape of an ever-changing city.

In conversation with insights from Taipei in the chapter by Sugimoto, the chapter by Erik Harms[29] describes the contentious ecological histories of urbanization and the environment in Saigon. In this case, we are invited to trace cycles of ecological vitality and withering as wetlands, lakes, and other water bodies were submerged to create new city building stock. This chapter offers a unique window into the ways that an active public sphere made up of architects, activists, and planners worked to create and critique massive new urban zones in Saigon. As in Sugimoto's study, Harms' analysis highlights the ways that new environmental and social agendas brought fresh forms of social exclusion to the city landscape. The chapter shows how landscape changes that were broadly understood as ecological improvements activated new modalities of living, thriving, withering, and loss: fresh boundaries between people and changed mosaics of urban nature.

Turning to the ways that infrastructures animate the urban ecologies of cities, Harris Solomon[30] looks more closely at the life cycles of city infrastructure itself.[31] In a case from Mumbai, Solomon considers the seasonal, deadly, and often hidden holes that make up the city's topography of potholes. The city's pothole-ridden roadscape returns with frequent storms and floods, reminding us how ecological cycles as fundamental as water and weather can assume an active role in lived cycles of urban risk, accident, and injury. Through this case study, Solomon shows how

---

26. For an example of such processes from south India, see Ishii (2017), which describes the worship of fierce gods and spirits angered and displaced by the construction of a major petrochemical complex and the development of a special economic zone in Mangalore.
27. See Tomonori Sugimoto, "The Death and Life of Urban Ecological Commons in Taipei," chapter 6 in this volume.
28. For an account of the struggle over a tiny enclave in old Bangkok, where poor residents mobilize claims of care and ritual belonging, see Herzfeld (2016).
29. See Erik Harms, "Concrete Ecology: Covering and Discovering Saigon's Ecology in a Time of Floods," chapter 8 in this volume.
30. See Harris Solomon, "The Absent Presence: Potholes in Urban India," chapter 5 in this volume.
31. An exemplary study of infrastructure and civic life in Mumbai is provided by Anand (2017).

seasonal floods and the potholes they entail can unleash a complex civic politics of urban nature.

If city floods draw our focus toward excessive rains, concrete, and impermeability, a contribution from Shubhra Gururani[32] offers contrast through her study of water that has disappeared. She considers the "embedded ecologies" of Gurgaon that no amount of planning and property-making can entirely erase. By describing the Ghata Jheel, an old lake in Gurugram, she shows how Gurgaon became a profoundly "unnatural" global city that lacks drainage, sewerage, and water supply. Invested as it is in business from knowledge industries and their attendant green imaginaries, the Gurugram in Gururani's analysis is hard pressed to present a vital ecology or a sustainable face.

Gururani uses this case in part to illuminate the hubris of development: planners expect to replace the complex web of natural and purposefully built water bodies with new, large lakes that are chiefly ornamental in purpose. Their prior function—enabling past urban settlement and agrarian practice—is almost entirely erased. In other contexts, cities in Asia have seen hasty and environmentally destructive development that has caused buildings to crumble on shaky wetland foundations. In this sense, wetlands can become polyvalent signifiers of ecologies of urbanism.

In an era of urban greening and growing environmentalist sentiment in cities, wetlands can simultaneously become targets for habitat conservation and sites for new urban sprawl, or locations for the displacement of the urban poor into resettlement colonies.[33] Across India, they embody the struggle for life and death in which nature in various forms is wrapped across cities. As Neha Sinha writes, "how India negotiates the distinction of wetlands—rather than homogenizing wetland as land, or protecting it only as forest—will set the tone for the future of wetlands in the country" (Sinha 2019, 152).

Turning to the death and life of urban nature in distress, a contribution from Kasia Paprocki[34] reminds us that contemporary cities are often also on the move—in this case, movement by ecological refugees. In her contribution, climate change is the main driver of policy and suffering in areas like the mangrove forested coastal villages of the Sunderbans in India and Bangladesh. By looking closely at climate adaptation regimes in these highly vulnerable and starkly poor areas, Paprocki problematizes the ways that evacuation has come to be the only sustainable and sensible response to climate change in the area. Paprocki's chapter illuminates how the weight of history, and a growing incapacity of states and international agencies to imagine a viable future for coastal rural areas, has real consequences for Khulna. It also turns our attention back to the importance of the scale of our analytical lens:

---

32. See Shubhra Gururani, "Making Land Out of Water: Ecologies of Urbanism, Property, and Loss," chapter 7 in this volume.
33. For a fine example of such processes, see Coelho and Raman (2013).
34. See Kasia Paprocki, "The Village at the End of the World: Ecologies of Urbanism in Climate Crisis Imaginaries," chapter 3 in this volume.

as localized experiences of climate change give way to new patterns of dislocation and migration, ideas and practices of nature-making—not to mention experiences of the loss of certain forms of nature—diffuse in new ways.

Paprocki's analysis invites this rendering of an ecology of urbanism into conversation with scholars such as Sunil Amrith, whose work reminds us that Asia, and particularly the eastern coastal edge of South Asia, is highly at risk because of its acute social inequalities. As he has written, "warming seas meet coastal zones that sag under the weight of growing cities, many of them founded as colonial ports in the eighteenth and nineteenth centuries. River deltas are sinking, starved of sediment by large dams upstream that were built in the 1950s and 1960s. We live with the unintended consequences," he writes, "of earlier generations' dreams and fears of water" (Amrith 2018, 5).

Another set of chapters elucidate projects of urban greening and the ways that benign and beautiful nature is invited back into the city or installed with renewed affection for experiences lost in industrial townships. In these cases, we are introduced to practices of gardening and farming, where sometimes rich legacies of urban gardens are recast as new projects of sustainable urbanism. At the risk of oversimplification, we might think of modern urban gardens as existing in two broad domains: public and private. The former, often manicured and heavily managed, tended to exhibit the features of botanical gardens that emerged out of the massive colonial exchange and study of flowering plants and trees that began in the early nineteenth century (S. Srinivas 2015, 57).[35]

Private gardens, on the other hand, were often less ornate and more functional but were crucial both to food provision and to the possibility for direct access to the natural world. Precolonial horticultural gardens in Bengaluru, for instance, were interspersed across the city, linked to water bodies like tanks and canals. They supplied fresh fruits, flowers, and vegetables. Today, their remnants appear in the form of nurseries and small vegetable gardens that feed urban residents. These are intermingled with other gardens that are associated with temples and shrines, which nourish the spirit and religious sentiments of modern city folk (S. Srinivas 2015, 52–64). However, public and private gardens were mutually imbricated in design, use, and aesthetic in many Asian cities as a legacy of colonial botanical gardens.[36]

In addition, many Asian cities are dotted with theme parks, civic parks, and what might be called accidental green spaces—including overgrown plots, unused land, and abandoned sites. There are also zoological parks in many Asian cities.[37] In many ways, earlier, modern Asian cities encountered urban nature as a spectacle, a playground, a food source, a symbol of spiritual power, and as a legitimizing source

---

35. A longer history of these biotic transplants and exchanges can be found in several key studies including Crosby (2004); Schiebinger (2007); Schiebinger and Swan (2007).
36. For an elaboration of this point, see the article by Besky and Padwe (2016, 17–19).
37. For a fine study of an Asian city zoo, see Miller (2013).

for political authority. Many of these aspects remain true today, but contributions to this volume explore ways these are transforming.

For example, it is the notion of gardens and political authority that comes to the fore in the chapter by Annu Jalais.[38] Considering the case of Singapore, she asks: what does it mean to declare Singapore a garden city?[39] Even as the total city area under various kinds of green and open space shrank in Singapore in the last decades of the twentieth century, city dwellers became more engaged in nature conservation issues. The chapter considers several processes, including control and culling measures directed toward crows and cats, while at the same time urban development brings spectacular gardens and agricultural parks to life.

Others have considered the greening of Singapore a political project of an autocratic government, or a particular response to the anxieties caused by Anthropocene awareness (Heejin 2017; Schneider-Mayerson 2017). Jalais, on the other hand, shows the ways that government projects to establish specific kinds of urban nature are also educational and discipline initiatives enacted with the city's public in mind. A growing and diverse awareness of urban nature, which she tracks in part through exercises associated with a group of students she teaches, allows Jalais to demonstrate the inevitable escape of nature from within carefully curated gardens, even in a most orderly city like Singapore. A quest for a less regulated life joins, then, with plants, animals, birds, and young people, pondering the elegant fortress their island state creates.

Emergent socionatural relationships and new modes of political agency are also at the center of a contribution from Camille Frazier.[40] Using edible gardens as a site for consideration in Bengaluru, she shows how her interlocutors respond to concerns about food safety and health by cultivating their own organic terrace gardens. In this way urban nature, as a regenerative and sustaining force, is brought into the domestic sphere, affording quite direct experiences of the rhythms and seasons of growing plants and vegetables. Families provide labor and coordination; this extends to both cooperation and conflict with neighbors. Frazier argues that direct farming in apartments creates new conditions to embrace a lively and life-giving nature in a context where the death of nature threatens the life of the city at large.

A third case explores issues of cultivation in urban nature through a detailed study of a specialty organic restaurant and its foodshed in Hangzhou, China. Here, Caroline Merrifield[41] considers food safety scares and environmental movements that advocate for more sustainable food production and consumption in the city.

---

38. See Annu Jalais, "The Singapore 'Garden City': The Death and Life of Nature in an Asian City," chapter 4 in this volume.
39. There is a considerable literature on the development of Singapore as a garden city. For some notable examples, see Ho, Woon, and Ramdas (2013); key cartographic and other history considered by de Koninck (2008) and de Koninck (2017) provide a record of the physical transformation and urbanization of Singapore Island.
40. See Camille Frazier, "Putting the Garden Back: Cultivating Life through Urban Gardening in India," chapter 2 in this volume.
41. See Caroline Merrifield, "Keeping Pace with the Foodshed in Hangzhou," chapter 9 in this volume.

In this case, organic food production and marketing serve as modes for providing a social experience of connection to producers who, in the past, were obscured by previous norms of processing, packaging, and retailing food. Merrifield's case details the complex interplay of food production cultures and the logics of specific food preferences in Hangzhou; this affords a broader commentary on the ways that intimacy with agrarian environments becomes a means to embody and nourish ecologies of urbanism, and to construct a positive, generative view of urban nature in China.

Considered together, the cases assembled in this volume offer a robust and dynamic account of urban nature, across the entire continuum that spans the death and return to life of nature. They provide grounds for our collective assertion that while displacement and extinction signify the death of nature in urban environments, cycles of life are not entirely broken. Rather, they are often re-organized in careful interplay with transforming social conceptualizations of what belongs, and doesn't belong, in the socionatural complex that is the twenty-first century Asian city. Diverse forms of urban nature can and do return to cities uninvited, but cases offered here remind us that they also return through complex social acts of solicitation, invitation, and ideas of communion and replenishment. Such ecologies of urbanism are always in motion across Asian cities, large and small.

When understood as multiple and nested urban natures, the analytical approach our contributors take provides a fuller sense of how sociocultural realms of ethics, aesthetics, and inhabitation—human and nonhuman—figure powerfully in city ecologies. It is here, after all, in the multiple and multivalent assemblages through which urbanism may be "greened" or made more sustainable, that the aesthetic and moral universes in which urbanism emerges also encompass social ideas about meaningful and vital life. Our hope in this collection is that each case amplifies how the twenty-first century Asian city environment can encompass both life and afterlife in the face of new social pressures and expectations. In this sense, the death of nature just might also occasion new imaginaries of lively and life-giving nature *in* and *for* the city.

Works Cited

Amrith, Sunil. 2018. *Unruly Waters: How Rains, Rivers, Coasts, and Seas Have Shaped Asia's History*. New York: Basic Books.

Anand, Nikhil. 2017. *Hydraulic City: Water and the Infrastructures of Citizenship in Mumbai*. Durham, NC: Duke University Press.

Baviskar, Amita. 2015. "Cows, Cars and Cycle Rickshaws: Bourgeois Environmentalists and the Battle for Delhi Streets." In *Elite and Everyman: The Cultural Politics of the Indian Middle Class*, edited by Amita Baviskar and Raka Ray, 391–418. Delhi: Routledge.

Baviskar, Amita. 2018. "City Limits: Looking for Environment and Justice in the Urban Context." In *Rethinking Environmentalism: Linking Justice, Sustainability, and Diversity*, edited by Sharachchandra Lele, E. S. Brondizio, J. Byrne, G. M. Mace, and J. Martinez-Alier, 85–97. Cambridge, MA: MIT Press.

Besky, Sarah, and Jonathan Padwe. 2016. "Placing Plants in Territory." *Environment and Society: Advances in Research* 7: 9–28.

Bhattacharyya, Debjani. 2018. *Empire and Ecology in the Bengal Delta: The Making of Calcutta*. Cambridge: Cambridge University Press.

Bonneuil, Christophe, and Jean-Baptiste Fressoz. 2016. *The Shock of the Anthropocene: The Earth, History and Us*. Translated by David Fernbach. London: Verso.

Brown, Frederick. 2016. *The City Is More Than Human: An Animal History of Seattle*. Seattle: University of Washington Press.

Carson, Rachel. 1962. *Silent Spring*. New York: Houghton Mifflin.

Chakrabarty, Dipesh. 2018. "Anthropocene Time." *History and Theory* 57 (1): 5–32.

Charlop-Powers, Sarah. 2016. "How Can We Make Urban Nature and Its Value More Apparent, More 'Visible' to People?" *The Nature of Cities*. September 2016. Accessed February 20, 2017. www.thenatureofcities.com.

Cioc, Mark. 2006. *The Rhine: An Eco-biography, 1815–2000*. Seattle: University of Washington Press.

Clifford, Mark. 2019. *The Greening of Asia: The Business Case for Solving Asia's Environmental Emergency*. New York: Columbia University Press.

Coelho, Karen, and Nithya Raman. 2013. "From the Frying Pan to the Floodplain: Negotiating Land, Water and Fire in Chennai's Development." In *Ecologies of Urbanism in India: Metropolitan Civility and Sustainability*, edited by Anne Rademacher and K. Sivaramakrishnan, 145–68. Hong Kong: Hong Kong University Press.

Crosby, Alfred. 2004. *Ecological Imperialism: The Biological Expansion of Europe, 900–1900*. Cambridge: Cambridge University Press.

Crutzen, Paul J. 2002. "Geology of Mankind." *Nature* 415 (6867): 23.

Crutzen, Paul J., and Eugene. F. Stoermer. 2000. "The 'Anthropocene.'" *Global Change Newsletter* 41: 17–18.

Dasgupta, Rajib. 2012. *Urbanising Cholera: The Social Determinants of Its Re-emergence*. Delhi: Orient Blackswan.

de Koninck, Rodolph. 2008. *Singapore: An Atlas of Perpetual Territorial Transformation*. Singapore: National University of Singapore Press.

de Koninck, Rodolph. 2017. *Singapore's Permanent Territorial Revolution: Fifty Years in Fifty Maps*. Singapore: National University of Singapore Press.

Dixon, Terrell, ed. 2002. *City Wilds: Essays and Stories about Urban Nature*. Athens: University of Georgia Press.

Doron, Assa, and Robin Jeffrey. 2018. *Waste of a Nation: Garbage and Growth in India*. Cambridge, MA: Harvard University Press.

Duara, Prasenjit. 2014. *The Crisis of Global Modernity: Asian Traditions and a Sustainable Future*. Cambridge: Cambridge University Press.

Elverskog, Johan. 2020. *The Buddha's Footprint: An Environmental History of Asia*. Philadelphia: University of Pennsylvania Press.

Fuller, Christopher. 2003. *The Renewal of the Priesthood: Modernity and Traditionalism in a South Indian Temple*. Princeton, NJ: Princeton University Press.

Gajjar, Sumetee. 2016. "Resilience through Caring for Nature in Times of Transition." *The Nature of Cities*. November 2016. Accessed February 20, 2017. www.thenatureofcities.com.

Gandhi, Ajay. 2012. "Catch Me If You Can: Monkey Capture in Delhi." *Ethnography* 13 (1): 43–56.

Gandy, Matthew. 2014. *The Fabric of Space: Water, Modernity and the Urban Imagination*. Cambridge, MA: MIT Press.

Garrard, Greg, Gary Handwerk, and Sabine Wilke. 2014. "Introduction: 'Imagining Anew: Challenges of Representing the Anthropocene.'" *Environmental Humanities* 5: 149–53.

Ghertner, Asher. 2015. *Rule by Aesthetics: World Class City Making in Delhi*. New York: Oxford University Press.

Ghosh, Amitav. 2017. *The Great Derangement: Climate Change and the Unthinkable*. Chicago: University of Chicago Press.

Grove, J. Morgan, Mary L. Cadenasso, Steward T. A. Pickett, Gary E. Machlis, and William R. Burch Jr. 2015. *The Baltimore School of Urban Ecology: Space, Scale and Time for the Study of Cities*. New Haven, CT: Yale University Press.

Guha, Ramachandra, and Juan Martínez-Alier. 1997. *Varieties of Environmentalism: Essays North and South*. London: Earthscan.

Hallman, Caspar et al. 2017. "More Than 75 Percent Decline over 27 Years in Total Flying Insect Biomass in Protected Areas." *PLoS ONE* 12 (10): 1–21.

Haraway, Donna. 2003. *The Companion Species Manifesto: Dogs, People, and Significant Otherness*. Chicago: Prickly Paradigm Press.

Hardach, Sophie. 2015. "How the River Thames Was Brought Back from the Dead." BBC, November 12, 2015. Accessed May 10, 2020. http://www.bbc.com/earth/story/20151111-how-the-river-thames-was-brought-back-from-the-dead?ocid=ww.social.link.email.

Harms, Erik. 2016. *Luxury and Rubble: Civility and Dispossession in the New Saigon*. Berkeley: University of California Press.

Harvey, David. 1996. *Justice, Nature and the Geography of Difference*. Oxford: Blackwell Publishers.

Heejin, Han. 2017. "Singapore, a Garden City: Authoritarian Environmentalism in a Developmental State." *The Journal of Environment & Development* 26 (1): 3–25.

Herzfeld, Michael. 2016. *Siege of the Spirits: Community and Polity in Bangkok*. Chicago: University of Chicago Press.

Hoffman, Richard. 2007. "Footprint Metaphor and Metabolic Realities: Environmental Impacts of Medieval European Cities." In *Natures Past: The Environment and Human History*, edited by Paolo Squatriti, 288–325. Ann Arbor: University of Michigan Press.

Ho, Elaine Lynn-Ee, Chih Yuan Woon, and Kamalini Ramdas, eds. 2013. *Changing Landscapes of Singapore*. Singapore: NUS Press.

Ishii, Miho. 2017. "Caring for Divine Infrastructures: Nature and Spirits in a Special Economic Zone in India." *Ethnos* 82 (4): 690–710.

Jensen, Casper Bruun. 2017. "Pipe Dreams: Sewage Infrastructure and Activity Trails in Phnom Penh." *Ethnos* 82 (4): 627–47.

Johnson, Nathanael. 2016. *Unseen City: The Majesty of Pigeons, the Discreet Charm of Snails & Other Wonders of the Urban Wilderness*. Danvers, MA: Rodale.

Kaika, Maria. 2004. *City of Flows: Modernity, Nature and the City*. London: Routledge.

Kelly, Matthew. 2018. "The Thames Barrier: Climate Change, Shipping and the Transition to a New Envirotechnical Regime." In *Histories of Technology, the Environment and Modern Britain*, edited by Jon Agar and Jacob Ward, 206–30. London: UCL Press.

Kolbert, Elizabeth. 2015. *The Sixth Extinction: An Unnatural History*. New York: Picador.

Krishen, Pradeep. 2006. *Trees of Delhi: A Field Guide*. London: Dorling Kindersley.

Latour, Bruno. 2013. *An Inquiry into Modes of Existence: An Anthropology of the Moderns*. Translated by Catherine Porter. Cambridge, MA: Harvard University Press.

Lefebvre, Henri. (1970) 2003. *The Urban Revolution*. Minneapolis: University of Minnesota Press.

Malm, Andreas, and Alf Hornborg. 2014. "The Geology of Mankind? A Critique of the Anthropocene Narrative." *The Anthropocene Review* 1 (1): 62–69.

Martínez-Alier, Juan. 2002. *The Environmentalism of the Poor: A Study of Ecological Cconflicts and Valuation*. Cheltenham: Edward Elgar.

McKibben, Bill. 2006. *The End of Nature*. New York: Random House.

McShane, Clay, and Joel Tarr. 2007. *The Horse in the City: Living Machines in the Nineteenth Century*. Baltimore, MD: Johns Hopkins University Press.

Merchant, Carolyn. 1990. *The Death of Nature: Women, Ecology, and the Scientific Revolution*. New York: Harper One.

Miller, Ian J. 2013. *The Nature of the Beasts: Empire and Exhibition at the Tokyo Imperial Zoo*. Berkeley: University of California Press.

Morrison, Kathleen. 2015. "Provincializing the Anthropocene." *Seminar* 673: 75–80.

Mullin, Molly, and Rebecca Cassidy, eds. 2007. *Where the Wild Things are Now: Domestication Reconsidered*. Oxford: Berg.

Nagendra, Harini. 2016. *Nature in the City: Bengaluru in the Past, Present and Future*. Delhi: Oxford University Press.

Oppenheimer, Michael et al. 2019. "Sea Level Rise and Implications for Low-Lying Islands, Coasts and Communities." In *IPCC Special Report on the Ocean and Cryosphere in a Changing Climate*, edited by Pörtner, H. O. et al. Accessed June 2, 2020. https://www.ipcc.ch/site/assets/uploads/sites/3/2019/11/08_SROCC_Ch04_FINAL.pdf.

Peet, Richard, and Michael Watts, eds. 2004. *Liberation Ecologies: Environment, Development and Social Movements*. New York: Routledge.

Piketty, Thomas. 2014. *Capital in the Twenty-first Century*. Translated by Arthur Goldhammer. Cambridge, MA: Harvard University Press.

Rademacher, Anne. 2017. *Building Green: Environmental Architects and the Struggle for Sustainability in Mumbai*. Berkeley: University of California Press.

Rademacher, Anne, Mary L. Cadenasso, and Steward T. A. Pickett. 2019. "From Feedbacks to Coproduction: Toward an Integrated Conceptual Framework for Urban Ecosystems." *Urban Ecosystems* 22 (1): 65–76.

Ramesh, Niranjana. 2018. "Infrastructures with a Pinch of Salt: Comparative Techno-politics of Desalination in Chennai and London." Unpublished PhD thesis. University College London.

Ray, Sugata. 2019. *Climate Change and the Art of Devotion: Geoaesthetics in the Land of Krishna, 1550–1850*. Seattle: University of Washington Press.

Rose, Deborah Bird, Thom van Dooren, Matthew Chrulew, Stuart Cooke, Matthew Kearnes, and Emily O'Gormand. 2012. "Thinking through the Environment, Unsettling the Humanities." *Environmental Humanities* 1: 1–5.

Schiebinger, Londa. 2007. *Plants and Empire: Colonial Bioprospecting in the Atlantic World*. Cambridge, MA: Harvard University Press.

Schiebinger, Londa, and Claudia Swan, eds. 2007. *Colonial Botany: Science, Commerce, and Politics in the Early Modern World*. Philadelphia: University of Pennsylvania Press.

Schneider-Mayerson, Matthew. 2017. "Some Islands Will Rise: Singapore in the Anthropocene." *Resilience: A Journal of the Environmental Humanities* 4 (2–3): 166–84.

Schwenkel, Christina. 2020. *Building Socialism: The Afterlife of East German Architecture in Urban Vietnam*. Durham, NC: Duke University Press.

Searle, Llerena. 2016. *Landscapes of Accumulation: Real Estate and the Neoliberal Imagination in Contemporary India*. Chicago: University of Chicago Press.

Sen, Sudipta. 2019. *Ganges: The Many Pasts of an Indian River*. New Haven, CT: Yale University Press.

Sharma, Sameer. 2018. *Smart Cities Unbundled: Ideas and Practices of Smart Cities in India*. New Delhi: Bloomsbury India.

Shelton, Barrie, Justyna Karakiewicz, and Thomas Kvan, 2010. *The Making of Hong Kong: From Vertical to Volumetric*. London: Routledge.

Sinha, Neha. 2019. "Water under the Bridge: Wetland Use and Abuse in India." In *Nature Conservation in the New Economy: People, Wildlife and the Law in India*, edited by Ghazala Shahabuddin and K. Sivaramakrishnan, 135–58. New Delhi: Orient Blackswan.

Sivaramakrishnan, K. 2015. "Ethics of Nature in Indian Environmental History." *Modern Asian Studies* 49 (4): 1261–310.

Sivaramakrishnan, K. 2017. "Courts, Public Cultures of Legality, and Urban Ecological Imagination in Delhi." In *Places of Nature in Ecologies of Urbanism*, edited by Anne Rademacher and K. Sivaramakrishnan, 137–61. Hong Kong: Hong Kong University Press.

Sivaramakrishnan, K. 2019. "Foreword." In *Caring for Glaciers: Land, Animals, and Humanity in the Himalayas*, edited by Karine Gagne, ix–xi. Seattle: University of Washington Press.

Srinivas, Smriti. 2015. *A Place for Utopia: Urban Designs from South Asia*. Seattle: University of Washington Press.

Srinivas, Tulasi. 2018. *The Cow in the Elevator: An Anthropology of Wonder*. Durham, NC: Duke University Press.

Steffen, Will, Jacques Grinevald, Paul Crutzen, and John McNeill. 2011. "The Anthropocene: Conceptual and Historical Perspectives." *Philosophical Transactions of the Royal Society* 369 (1938): 843.

Tarlo, Emma. 2003. *Unsettling Memories: Narratives of the Emergency in Delhi*. Berkeley: University of California Press.

Tsing, Anna. 2015. *The Mushroom at the End of the World: On the Possibility of Life in Capitalist Ruins*. Princeton, NJ: Princeton University Press.

Tsing, Anna, and Nils Bubandt. 2018. "Feral Dynamics of Post-Industrial Ruin: An Introduction." *Journal of Ethnobiology* 38 (1): 1–7.

Tsing, Anna, Heather Swanson, Elaine Gan, and Nils Bubandt, eds. 2017. *Arts of Living on a Damaged Planet: Ghosts and Monsters of the Anthropocene*. London: University of Minnesota Press.

Tsing, Anna, Jennifer Deger, Alder Keleman-Saxena, and Feifei Zhou, eds. 2020. *Feral Atlas: The More-than-Human Anthropocene, Stanford Digital Projects*. Stanford, CA: Stanford University Press.

Van Horn, Gavin and Dave Aftandilian, eds. 2015. *City Creatures: Animal Encounters in the Chicago Wilderness*. Chicago: University of Chicago Press.
Walker, Brett. 2011. *Toxic Archipelago: A History of Industrial Disease in Japan*. Seattle: University of Washington Press.
Warde, Paul. 2011. "The Invention of Sustainability." *Modern Intellectual History* 8: 153–70.
Warde, Paul. 2018. *The Invention of Sustainability: Nature and Destiny, c 1500–1870*. Cambridge: Cambridge University Press.
Waters, C. Jan Zalasiewicz, C. P. Summerhayes, and Anthony Banrnovsky. 2016. "The Anthropocene Is Functionally and Stratigraphically Distinct from the Holocene." *Science* 351 (6269): 137.
Woolfson, Esther. 2014. *Field Notes from a Hidden City: An Urban Nature Diary*. London: Granta Books.
Zalasiewicz, Jan, Mark Williams, Alan Smith, Tiffany L. Barry, Angela L. Coe, Paul R. Bown, Patrick Brenchley, et al. 2008. "Are We Now Living in the Anthropocene?" *GSA Today* 18 (2): 4–8.

1

Divine Excess: The Power of Accidents and Nature Spirits in Bangkok

Andrew Alan Johnson

In October and November 2011, Bangkok flooded. A combination of climate change–boosted rainfall and an expansion of paved space in the Chao Phraya river valley meant that the yearly rainy season caused massive amounts of runoff—runoff that swept down the river toward the city. Seeking to protect the valuable downtown, the Bangkok Metropolitan Authority placed barriers around its most expensive real estate, shunting the water toward the city's outskirts and its poorer sections.

In one of these sections to the north of the city center, as the flood took a turn directly toward a neighborhood Buddhist temple, some of the local villagers and workmen gathered to protect the grounds. One man was working at the back of the temple, clearing some trees that were clinging to the bank and making ground on which sandbags could be laid. As he swung his machete to chop a *tani* banana tree (*Musa balbisiana*, an inedible ancestor of the common banana), the machete broke and the blade flew into the floodwaters. The man returned with a new machete and the exact same thing happened. The third time the machete broke, as he watched, he saw a female shape form in the pattern of leaves and stalks of the tree, then disappear.

"This is how we knew it was this tree," a monk at the temple told me a year later. The tree, the monk indicated, was not simply a tree. Rather, it had living within it a female spirit [*nang tani*] that had emerged in the wake of disaster. The newspaper came to report on the visitation, and the temple was flooded again—this time with visitors seeking to ask things of this banana tree growing incongruously in the city. The temple built a shrine to the banana tree's spirit, as both her association with misfortunes big (the flood) and small (the serial destruction of the machete) were testaments to her power.

But this was no solitary occurrence. Just south of the temple, on the road heading into Bangkok's center, and just as the center traffic island widens near the entrance to Bangkok's Criminal Court, are row after row of wooden statues of zebras and giraffes, facing outward toward the oncoming traffic. Their white-and-black stripes (and, for the giraffes, yellow-and-black reticulation) mirror the white, black, and yellow of the road surface. Their wooden figures are donated as gifts to beings like the banana spirit, spirits that that dwell at sites of accident and disaster. While they can be the cause of accident, these spirits, ghosts, and other uncanny beings—often the specters of animals or plants whose deaths allowed the expansion of Bangkok's urban sprawl—also offer lottery numbers, healing, protection, and belonging to those urban residents passing through or living in their spaces precisely *because* of their legacy of accident. The beings, wearing the colors of the highway and the forms of animals, unify the thought-to-be-overcome animal world with the planner's logic of the road system.

These shrines proliferate in Bangkok's industrial halo—the ring around the city where factories employ day laborers and where informal-economy workers live. On the western edge of the city, there is a shrine to a lady mother King Cobra, a snake killed during the construction of Rama II Road, the multilane artery that takes traffic to Thailand's South. On lively Ratchadapisek road, to the north of the city center, a tree trunk lays in a temple where it has been dredged out of the Chao Phraya river, and its possessing lady mother is venerated as a source of both power and menace to those seeking it out. In Bangkok's working-class east, the spirit of a woman killed in childbirth also looks after an ancient, capsized boat pulled from a neighboring canal, along with giant hardwood tree trunks discovered buried during construction.

These spaces provide a kind of mirror to rationalized green space in the city centers, where architects place their silhouettes of *flaneurs*. The two are in tension, as the city seeks to demolish the overgrown vacant lots where such shrines proliferate in favor of a green, rationalized city (see Harms 2012 for the irony of such efforts). These latter efforts, lauded by middle-class Bangkokians, hold as a model European or East Asian urban green spaces, especially Singapore (Johnson 2015). Such efforts involve corralling street vendors into particular zones, off the sidewalk, and a bicycle-share program that encouraged residents to cycle through Bangkok's notoriously dangerous traffic—further complicated by the fact that those few who could afford to live in zones with the kiosks often had their own drivers. Similarly, Bangkok's plastic bag ban was aimed at middle-class consumers and recruited large-scale department stores—places where the majority of Thais cannot afford to shop (Methanuphab 2019).

Here, environmentalism is a hobby of the well-off, and environmentalist concerns become an aspiration of the rich, a moral quest linked with cosmopolitanism and monarchy—a combination that in Thailand shares a link (see Peleggi 2002). In the city's imaginary environmentalism is a part of the high-modern becoming of the city, rather than a process of understanding the ways of life that this becoming erases. Thus, the spirit inhabiting urban green spaces reflects modernist aspirations and an embeddedness within global/civilizational hierarchies of value. There is aspiration in such green spaces; not only the mourning, trauma and loss that I discuss here but also a fetishization of "progress" haunted/blessed by the selectively edited histories upon which these sites are built (see also Johnson 2014). These efforts appealed to a global notion of eco-friendly life, one that drew upon an image of Bangkokian Thais (with the monarch as their head) as internationally linked cosmopolitans, as well as the long tradition of noblesse oblige that many rich and/or royal Bangkokians had cultivated (see Winichakul 2000).

But many of these green projects eventually failed. The bike-share program, "pun pun," simply left concrete barriers across sidewalks as unused bikes rusted in their cradles, further impeding foot traffic, and the much-heralded "monkey cheeks" project touted by Thailand's late king Bhumibol Adulyadej, designed to hold water in green zones during flood season, failed so spectacularly that, in the wake of the disastrous 2011 floods, Thai magazines had to scrub any reference to it.

As a detailed example of the reshaping of the city in the name of a modernist urban vision of green space, Michael Herzfeld (2016) describes Pom Mahakan, a community in the heart of Bangkok's Rattanakosin district. It was a vibrant and old community, shaded by hundreds-of-years-old trees and, from the air, appeared as a dense green rectangle in the gray Bangkok landscape. But, while the community was green in the sense of having thick tree cover, the Bangkok Metropolitan Administration (BMA) favored a green of an open grass lawn—a lawn that was to become the community's eventual fate, despite a decades-long standoff (see Harms 2012 for a Vietnamese example of such almost willful misunderstanding). While the BMA blamed the community for generating trash, the lawn that replaced the community became an informal dumping ground. Nature in the city, in other words, needed to come on planners' terms.

In contrast, the green spaces that I describe present particular challenges to the notion of sovereignty and infrastructure, accident and luck, green space and urban planning, and nature and culture. As authors to otherwise author-less accident, the spirits haunting Bangkok's nature shrines act in the present as a kind of divine excess. Here, by "divine excess," I mean that they are something in addition to the built environment, a sacralized ecology that serves not only as a point of critique and contradiction to the city's seemingly developed facade but also as a place of possibility for those struggling with life in Bangkok. Here, I trace the (re)emergence of tree spirits through the Bangkok concrete via an engagement with the literature on materiality and the city, and an extended discussion of one such shrine, at the site of a deadly fire in eastern Bangkok.

The Forest in the City

Bestiaries of such spirits are popular in English and in Thai (see Khamjan 2008; Textor 1973). But such catalogues are misleading in their presentation of a fixed Thai folklore. New spirits appear often, and new qualities become attributed to old spirits. Indeed, the discussions of such spirits in popular media give rise to sightings of similar ghosts—after the popularity of Japanese horror films, there was a resultant explosion in sightings of ghostly girls with long black hair, and reports of floating-head ghosts [*krasue*] proliferated following a popular Thai film (see Baumann 2014).

Here, the category of "spirit" (*phi*) itself becomes a kind of a catch-all for things beyond immediate perception. I see the excess that my interlocutors locate within ghostly beings as shifting, occasionally taking particular recognizable forms, but rarely remaining constant—in this I take inspiration from James Siegel's (2006) discussion of the figure of the witch as a "menace from all directions" (Siegel 2006, 171) that refuses to be named in post-Suharto Indonesia.

During my field research in 2012–2016, tree spirits [*nang mai*], especially those of the *tani* banana palm, were the ones to whom roadside shrines [*san jao*] were most commonly dedicated. If not *tani*, other shrines were dedicated to the larger tropical hardwood *takhian* (*Hopea odorata*) that also often houses a similar spirit, although my interlocutors spoke of *takhian* as more powerful and, for the most part, confined to the forest except when urban construction might unearth a preserved log (see Johnson 2012; 2016). I have, at other times, encountered shrines to *pradu* (*Pterocarpus macrocarpus*) or tamarind trees (*Tamarindus indica*). Finally, banyan trees (*Ficus benghalensis*), sacred in Hinduism and Buddhism, might also house ghosts—not spirits of the tree itself, but the ghosts of dead humans who had wandered in from elsewhere (*phi tai hoong*—see Johnson 2014).

These tree spirits share commonalities. They are, for the most part, beautiful women who are nonetheless dangerous, often trying to capture men's spirits for their own sexual pleasure via causing accidents. In spite of their danger, they might favor a person bringing gifts that flatter their vanity (dresses, combs, mirrors, etc.) with secret knowledge or assistance. But such attempts are dangerous, and the *nang mai* can take revenge when slighted.

Most of what I address here are *nang mai*. But there are other spirits in Bangkok. There are lordly beings associated with prestigious sites in the city, including *Phra Siam Thevarat*, the explicitly named "spirit of the nation" installed during the nineteenth century (see Jackson 2010); Hindu deities and shrines associated with them (see Ayuttacorn and Ferguson 2018; Keyes 2006); spirits with extremely specific portfolios, such as *jao hai-tek* (Lord High-Tech), who answers prayers directed toward technological devices; and there are, of course, the spirits of bad death, most of whom are unambiguously hostile but at least one of which, Ya Nak, fulfills a similar role to the *nang mai* I describe here (Johnson 2016).

Shrines to these above spirits, largely beings of the city and its residents, are more permanent. Ya Nak's shrine has stood in some form for nearly a century, for instance; Phra Siam's for half again as long. But the *nang mai* shrines appear and disappear quickly, on a timescale of weeks rather than months. Their devotees are more often migrants, themselves transitory in the city, and their existence is fueled (and undone) by rumor, social media posts, and speculation.

So—why women and trees? The city spirits (e.g., Phra Siam, Jao Hai-Tek, and others, with the exception of Ya Nak) of Bangkok are generally male, and nature spirits are largely female. Their devotees often refer to them as "mother" (*mae*) (see Johnson 2016). When I asked devotees at spirit shrines why tree spirits were always female, they would often point to the rippling effect of the wind through branches and leaves as evoking a woman's flowing hair. But a structuralist explanation might point toward a complementary opposition to the masculine power of Buddhism (only men may become monks) and Hindu kingship (while there are exceptions in regional history—see Swearer and Premchit 1998, most rulers are men). Most spirit devotion is held at night, while monks sleep. Gifts from spirits are often lottery numbers, or promises of a lover's devotion, or other issues with which monks should (at least in theory) not concern themselves. In this (admittedly over-simplified) explanation, nature, women, night, and personal concerns are juxtaposed with cities, men, kingship, daytime, and cosmic concerns. In this interpretation, the spirits of trees paved over by Bangkok's sprawl emerge as a barely repressed subaltern, a counter-discourse to Bangkok's high modernity.

But here, I seek to pursue a slightly different tack.

Cracks in the Concrete

It is little wonder that hyper-modern, planned concrete surfaces, especially roads, form a central part of these stories of tree spirits reemerging into urban space. It is large-scale concrete structures that are the principal concerns of Bangkok's urban planning: riverside bike paths, new rail lines, new mega shopping malls. Eli Elinoff (2017), in his analysis of Thai concrete, shows clearly the dual nature of concrete in Bangkok: it is both the means through which urban planning dreams take shape but also, via half-completed projects that lie in ruins in the city, the physical manifestations of failure and the focal point for accusations of corruption.

But it is not only fears of financial corruption that crawl through cracks in the concrete. In Jane Ferguson's analysis of the haunting of Bangkok's (then) brand-new Suvarnabhumi airport (2016), a massive edifice meant to showcase Thailand's place at the forefront of Southeast Asia, airport workers report seeing snakes, women in traditional clothing, and other ghostly manifestations in the high-modern space. For Ferguson, the manifestations are not necessarily indications of a desire for a premodern past underlying the modern present; instead, she argues, ghosts and aviation are both actors in the present. The destruction of Cobra Swamp [*nong ngu hao*] to make way for the airport does not mean that the eponymous snakes disappear. Rather, they are transformed and subsumed within the city, and they become a part of its urban fabric no less than the concrete runways—the two coexist.

So how should we conceptualize the emergent cities of ghostly women and snakes—that *alternative* green space—underneath the planned urban environment? Michel DeCerteau, in *The Practice of Everyday Life* (1984), asks us to consider two views of New York City. The first is from atop the World Trade Center, looking down upon the grid of streets below. This is the city of the planner, that nineteenth-century grid slashed with Broadway and ringed by Robert Moses' highways. For DeCerteau, to see the *other* New York, one must travel down to eye level and see the city as it is created in the everyday: the "certain strangeness that does not surface" (DeCerteau 1984, 93) into the map of the planner. As pedestrians choose their routes, and in doing so interact with others, they create another city, the *parole* to the planner's *langue*. They create a blind city defined by habit rather than plan, one composed of the impromptu entanglements that walkers have with others—the tree spirit shrine, as opposed to the green grassy park lawn.

But, as the emergence of ghostly presences indicates, there is no reason here to restrict ourselves to walking, nor to restrict ourselves to us. Forces beyond the human move and act in the city. Infrastructure and what it does has increasingly been a focus for new anthropological work: water and sewage run through pipes beneath the city, creating new paths for citizenship (Anand 2017); planners and residents offer contrasting conceptions of cool and hot air and the influence of the built environment (see Harms 2016); and urbanity fundamentally alters the light— and soundscape (Schivelbusch 1988).

But not all such entanglements are so peaceful. At times, as the presence of the tree spirits in Bangkok note, they can be disastrous. When we use the term "accident," we highlight the lack of human intentionality. Should a building catch fire, this intentionality is important. In English, arson is different from an accidental blaze, even if they have the same results. But the Thai *ubatihet* is more phenomenological than the English "accident." *Ubati*, a Pali word, means something more like "happening," a meaning reflected in *het*, "an event." Thus, an accident as *ubatihet* is simply "a thing that happened," with no claim toward intentionality (or lack of it). Indeed, as Trais Pearson (2020) shows, the term entered Thai within just over the past century, as a category of happening wherein the state held force, as opposed to interpersonal or intersubjective relationships. But, as Pearson shows, the idea of an *ubatihet* is more often than not abandoned in favor of a resolution that involves agentive actors, and not just "happenings." Accidents are never author-less.

In *Vibrant Matter*, Jane Bennett (2009) makes the case that nonhuman material is also an actor: rubber tires fail to grip wet road surfaces, rain makes holes in the pavement, motorbikes fall. As in DeCerteau's walking city, actants meet, entangle, and in so doing create new entanglements. Thus, taking a vibrant matter approach in which intentionality is secondary to action, every action is accident.

But this does not make for a satisfying story. In a deadly accident, justice calls that we find intentionality (or criminal inattention). Where does this intention lie? It is a problem that vexed Evans-Pritchard's Azande: an accident occurs and those who suffer from it seek an author but do not find one (Evans-Pritchard 1976 [1937]). For Evans-Pritchard, the search for the author of accident—the witch—is a search that exposes and diffuses tensions within Azande society. A victim wonders who harbors a grudge. And when that person is then confronted with the evidence, the witch must admit that his desire to harm, however unwitting, in fact inflicted it. But what to do when we bring the agentive nonliving into the mix? What is the social legacy of accident that truly has no human author, and where does the intentionality of this accident go?

In Bangkok, at the zebras that watch the street at places where accident has struck, this question of authorship is answered by the *jao*, the owners[1] of these sites of death. In the cases I discuss here, lady mothers [*jao mae*] associated with destroyed elements of the wilderness are both the authors of accident as well as sources where individuals can harness this excess for their own purposes. It is my argument that this divine excess arises from and forms a part of the urban ecology of Bangkok. Because of the unanswered question of the authorship of accident—unable to be dismissed as easily as Evans-Pritchard's witchcraft—such sites become infused with an excess, a sense of potency. In this unintended space, new possibilities emerge.

Spirits as Urban Ecology

But why do tree spirits demand statues of African animals whose colors evoke the road surface? In other words, what is the link between cosmology, infrastructure, and ecology?

It has become a matter of course to point out how nature and culture are inseparable. Hugh Raffles (2002) and Julie Cruikshank (2005), in two excellent studies of the social construction of nature set in the Amazon and the sub-Arctic, respectively, point out how an environment always in tension with human activity (be it social or ecological) becomes recast as untouched, primeval nature. In a parallel move, urban ecologists do not end their work where the planner's work begins, but rather they conceive of urban zones as functioning ecologies in their own right—the urban is just another biome.

With this interconnection of the urban and the ecological in mind, Anne Rademacher calls upon us to look at the process of "making the environment" (Rademacher 2011, 30), a perspective toward how socionatural (Rademacher 2011) spaces are historically imagined and valued specifically within the city. In this light, the recognition of Bangkok's lost greenery would point toward a religious recognition of this loss as a way of making green space in the city not the sterile monument to high modernity that spaces such as Pom Mahakan have become, but a living, intersubjective greenery.

But here, I argue, there is something more. What if we were to take the broad view on what constitutes a nonhuman presence in the city? I see urban ecology as the sum total of our entanglements, involving all the ontologically present

---

1. "Jao" can denote a noble [*jao nai* or simply *jao*], an owner of a thing [*jao khong*], or "spirit" in the sense that I use it here.

actors—beyond the animals, plants, and material of the built environment, but also beings of excess, such as spirits. I see Bangkok's shrines to dead trees, snakes, and other "natural" beings (see Johnson 2012) not as planned spaces for the urban dweller to interact with the natural world (e.g., as a sort of religiously inspired parkland), but rather in their unplanned, accidental capacity. Like DeCerteau's walking paths in the city, accidental entanglements undermine the planner's concept of the city, pointing to a potential of the city that exceeds and evades urban capture.

In so doing, I bring recent work on animism to the conversation on urban ecology. In recent years, anthropologists interested in the entangling of human and nonhuman worlds have revitalized the term. This work seeks to go beyond Tylorean notions of animism that see animism as a variety of belief to reflect instead the deep engagement that individuals have with the personhood of other beings (Århem 2016). Animists, in new animist studies, recognize the social agency of other beings, even of nonliving matter.

There is a natural synergy between ideas of vibrant matter such as Bennett's and animism (see Povinelli 2016 for a recent such analysis). And while most anthropologists of the new animism focus on small-scale societies in the Amazon or circumpolar regions, via welding the vibrancy of infrastructure with the agency of spirits, such an approach can easily be brought to bear upon the metropolis in Thailand. Trees and snakes in the forest speak, but so do asphalt and motorcycles in the city. Indeed, as I show here, the latter speak with the former's voices.

But here is a complication. Asphalt, like green space, already has a voice—from street signs to painted lines to traffic police. The city, unlike the forest, is a space intended to be already legible—indeed, it is this language of white and yellow lines on a black road surface that the colors of the zebras and giraffes given as offerings recall. Unlike the forest, one does not need to speculate about the hidden social owners of the intersection, because the owner is clearly delineated—the stamps on each manhole cover are marked with the BMA's seal. Thus, the arrival of new, other owners presents a case of divine excess, an idea that something exists beyond the planned environment. Indeed, this complication of sovereignty is reflected in the names for these spirits—*jao*, ruler, owner. This uneasy resting of (owned) urban space atop (already owned) natural space erupts in accident and death.

Ghosts of the City

The practice of devotion to urban guardian spirits [*phi meuang, jao meuang, phi arak*] has a long history in Thailand and other parts of Southeast Asia and rests upon a particular kind of division between civilized space and wilderness (see Johnson 2014; Århem 2016, 296–97 for other Southeast Asian examples). The village in which Stanley Tambiah worked, for instance, was divided between the domain of an ascetic, vegetarian ancestral spirit [*tapuban*—lit: "grandfather of the home"] and the wilderness of a meat-eating, more capricious spirit (1970). In the northern Thai city of Chiang Mai, rituals intended to recharge the city's fortunes depend upon the supplication of cannibalistic mountain spirits cowed into abeyance by the power of the Buddha and the peaceful spirits of the city (see Johnson 2014). Thus, wilderness spirits are in a productive division with those of the city. Once placed in their proper hierarchical position and their previously dangerous position nullified, they become sources of potency for the city, sources of rain, peace, and fortune.

There is, then, an additional city to the planner's grid—the city as a civilizing force, guided by divine presences. Cities in the medieval period mirrored the cosmic layout of the heavens, and in doing so, their planners sought to harness some of this divine power (Swearer and Premchit 1998). This divinity extended to their residents as well. Thongchai Winichakul describes the nineteenth-century construction of a hierarchy of residents in Siam (Thailand, before the 1930s), wherein colonial Europeans (and, at a previous time, Chinese) were seen to be of a higher rung, and Thai dwellers of the capital city [*chao krung*] were the gate through which European notions of modernity, development, and other benefits flowed to the rest of the populace (Winichakul 2000). Indeed, later nationalist constructions asserted that Bangkokians were uniquely able to combine European or Japanese technology with Buddhist morality in ways unavailable to either foreigners or country-dwellers [*chao bannok*] (Johnson 2014). Such constructions extend to the present—Daena Funahashi (2016) details how health experts in Bangkok made the case *against* democracy in Thailand by arguing that the rural poor existed on a differential (and lower) temporal, moral, and intellectual plane. Just as giving absolute freedom to a toddler would result in disaster, Thais without sufficient access to this moral and international force should be guided by those with. And here is where urban efforts to render the city green turn a blind eye toward magico-religious forms of making green space.

But the tree spirits here present a problem—spirits such as the banana tree ghost are not these civilizing forces of urbanity, like Phra Siam or like the green space that succeeded the community of Pom Mahakan. And, unlike in Chiang Mai, the return of nature spirits in urban Bangkok is not a peaceful one. It is not the happy return of a subjugated foe but the irruption of the thought-to-be-overcome into the everyday. It is an overturning of the established order, a violation of the city plan by what lies beneath the concrete. It is to this irruption that I now turn.

The Trees of Bangkok

In an overgrown vacant lot in the Ekkamai neighborhood of Bangkok, a large tamarind tree stands, festooned with dresses on hangers like a strange kind of Christmas tree (Figure 1.1). These dresses are traditional outfits but are made of cheap silk and not really designed to be worn. Rather, as their green and yellow colors—matching the foliage—indicate, they are gifts to something that cannot physically wear the dresses, but which can inhabit them anyway: the woman of the tree [*nang mai*].

Figure 1.1: Dresses for the tree spirit

Beneath them stands a cabinet where more dresses are hung, a mat for people to come and kneel in front of the tree with incense offerings, and a series of pencil-drawn portraits of a smiling woman. Also, of course, there are zebras.

But not every tree has a *nang mai*. And this lot is, for other reasons, special, as it is marked by death. Further back in the lot, fenced off to the public, is a cleared concrete space and piles of shapeless rubble, covered over in several years' worth of vegetation. These are the remains of the Santika Club nightclub, which burned down on New Year's Eve 2008–2009.

The lot is about two blocks away from being cool. Soi [lane] Ekkamai runs parallel to the super-trendy soi Thong Lor, the heart of Bangkok's nightlife scene for young and wealthy Thais. Where Santika stood was on the edges, ringed by residential zones and slum communities that ran down smaller pedestrian soi on either side of it.

On that New Year's Eve, a band (unfortunately named "Burn") was set to play at a party (also unfortunately titled "Santika's Last Night"). Something happened—investigators were unable to determine whether it was fireworks falling on the roof, pyrotechnics from the band itself, or an unrelated electrical fire—and a fire broke out. As the place burned, panicking patrons found that all exits but one were locked. Sixty-six people died in the blaze, and 222 suffered injuries.[2]

Reasons as to why the fire broke out varied. In 2012, I interviewed residents in the neighboring lanes and received conflicting responses. Some shooed me away, eager to move on. Some were wary of outsiders asking about the fire, as they had already been visited by threatening figures seeking to have them downplay the tragedy during the upcoming trial of the club owner. Others spoke of ghosts: "some people hear [the crackle and roar of] flames and screams at night," one older resident told me, "but I just close my eyes tight and try to sleep [when I hear them]." Others speculated on what might have given rise to the fire: the ground used to be a Muslim cemetery (Khao Sod 2009), it was the site of the grisly murder of a Navy officer (M-Thai 2012), the building was shaped like a coffin and even had a cross as decoration (Khao Sod 2009).[3]

Their stories are replete with various kinds of agency: neglectful or wicked club owners, the ghosts of ethnic and religious others (assumed to be more violent),[4] ghosts of the violently killed [*phi tai hoong*—see Johnson 2013], or even the power of a coffin's shape to bring forth death. But the residents hearing the ghosts at night were not those visiting the tree shrine. For the devotees to the tree spirit, the fire and its deaths stemmed from disrespect to the plants that grew up around it. According to these tellings, the tree stood next to the nightclub and had to endure loud music, car exhaust, and even the urine of drunken men in too much urgency to wait in the bathroom queue. Eventually, enough was enough, and the disrespected tree spirit burned the building down.

But those visiting were not those seeking to placate the spirit of the tree because they feared it—the tree spirit's thirst for vengeance had already been slaked. Nor were they relatives of the victims—those people found their own closure in criminal court. Instead, what those seeking out the tree told me was that they were looking to build a relationship with the spirit itself via an exchange of gifts: dresses for lottery numbers or love, zebras for protection, a hired dancer to perform a traditional dance [*ram*] for a new job (Figure 1.2).[5] They were looking for the potential that such accidents provided, to find a break in their own positionality within the city via the very same forces that had broken the seemingly planned club. Thus, the

---

2. The number of dead is yet another point of disturbing coincidence. Six in Thai [*hok*] rhymes with the word "to fall" [*tok*] and is considered unlucky.
3. Thailand's Christian community is tiny. Crosses, therefore, signify not Christianity but rather a Gothic-rock aesthetic, itself drawing upon themes of horror and death. A cross would give a room a faux Western funeral air, thus the association between the cross and the disaster.
4. In addition to long-standing prejudice against Muslims in Thailand, especially in the wake of the insurgency in Thailand's south, there is also a horror of graveyards. Buddhist Thais generally cremate, and historically graves were places where those violently killed and thus too unfortunate to be cremated would be interred.
5. These are examples; there is no set menu. The exact exchange is individually decided upon considering the magnitude of the favor. These exchanges are promised [*bon*] in advance and paid [*kae bon*] afterwards.

Figure 1.2: A dance as an offering

Santika fire was not entirely a result of human action but instead it was a crack in the everyday, exposing a potential beneath.

The cults of tree spirits in Bangkok are individual affairs. There is no one leader of a particular place. Rather, fame slowly spreads via word of mouth and social media posts, and people come based upon their own desires. This does naturally lend a certain rhythm to the place—on nights before the lottery, crowds gather—but no religious organization creates holy nights or particular forms of ritual assembly. But, as certain sites gain fame and crowds become difficult to manage (or potentially profitable), some figures do appear.

At one site near the Santika fire, also in eastern Bangkok, an eclectic group of mediums gathered along the road, featuring men engaged in numerology, a child who was said to be possessed by the tree spirit, and even a roti (fried bread) vendor whose dishes could sometimes reveal messages from the trees in the form of lucky numbers. The nightly crowds, during the month or so in 2012 when I visited this shrine, would form a circle around each expert, watching closely for any sign that some piece of wisdom was revealed.

While there are no official rolls of attendees, nearly everyone whom I met at these shrines were from Thailand's North or Northeast, internal migrant workers from the poorest parts of the country engaged in informal labor and often ethnically and linguistically different from Bangkokians.[6] Their labor could include construction, motorcycle driving, petty vending, and even sex work, but in each case, these were practices that involved an engagement with uncertainty (see Johnson 2012), and some degree of alienation from Bangkok's establishment.

Revelations from tree spirits at Santika or elsewhere came from dreams or patterns in the bark. Devotees spread oil on the trees, rubbing the trunk with their fingers until shapes appeared. The process was necessarily tactile. For instance, Jiap, a middle-aged devotee from the northern province of Phrae, would pour a bit of oil she'd bought from the vendor near the trees onto a bare space in the wood. She described her activity as "digging into numbers" [*khut lek*]. Shouldering open a space in front of the bark, she would rub her thumb in broad circles over the wood, already slick and shiny from others' attempts (Figure 1.3). After a few digs, she would look, frowning, at the spot, then try again or move. For others, cell phone cameras and their lights provided a second vantage point to examine the numbers that the tree might give. Some even used candles to throw these patterns into relief, others clucked their disapproval: "Fire!" Jiap said to me, "It burns her! She will give them wrong numbers if they keep doing that."

Jiap and others were more skeptical than I had anticipated. Popular depiction of spirit practices, seances, ghost hunters, and the like often show believers to be overly credulous, seeing spirit actions in every strange noise and shape. In the telling and retelling of uncanny tales at home in the United States, or in my previous fieldwork (Johnson 2014), pretending to see something in a photo is often part of a kind of speculative pleasure. *Might* there be a spectral face in the window? *Might* there be a sinuous shape under the water? But Jiap and her friends listened to my guesses on the shapes of numbers with mouths set in a straight line, responding "maybe" or "I don't see it."

But still they asked me. In this realm where the unusual and the strange mark the activity of spirits, the appearance of an unusual person also contains significance. When I began coming to Bangkok's shrines with a large camera, I was taken as such a sign. Here was a young white man, presumably wealthy (false), and—to top it all off—with what appeared to be an expensive camera (true). After a short period of time, I had a group of middle-aged women (including Jiap) following me, asking me to take pictures of the trees with my camera and examining the camera for numbers. When I protested that I would rather see *their* technique for finding numbers, they insisted I would be more inclined to find correct numbers. I was the unusual one, here, not them. For that reason, I was closer to the spirit—foreigners and spirits

---

6. Northern Thai (*kammeuang*) is its own language and had, up until the twentieth century, its own alphabet. Northeastern Thai is a dialect of Laotian, although many Northeasterners speak Khmer (Cambodian). Both *kammeuang* and Lao languages are related to, but not always intelligible with, Thai.

Figure 1.3: "Digging" for numbers

were naturally linked (Johnson 2020). Rather, they were seeking to *become* unusual (by winning the lottery, for instance). The uncovering of disaster—the tree, traffic accidents, a fire—means that something has penetrated the everyday. Proximity to that which breaks the everyday—the foreigner with the camera, for instance—meant that one could drink from that fountain of potential.

Jiap explained to me her technique for ensuring that the spirit gave correct numbers:

> The lady mother, she knows. She can see if you have fate [to be wealthy] (*chata*) or if you have merit [*bun*]. Things we do not know. She will help you more if you have more merit. She likes to help [virtuous] people like that. So what I do is when I win something, I give it to the temple. Or might use it to help out my family, but I make sure to give [a significant amount] to the temple. Then when I come back, she knows that I am a good person and might help me [more]. (Personal communication)

Here, Jiap presents a particular problem. She is not the kind of person who normally is lucky. She is, in a sense, stuck in her position as surely as the trees are stuck in concrete. But with the irruption of accident, she imagines then that her position might change. And were she savvy enough in negotiating the currents of Buddhist karma, cash, and fortune, she might be able to get out of this stuck position.

But she kept following me. "Why me?" I asked Jiap, as she asked me to take another photo. I had certainly been a worse Buddhist than she had been. She looked at me appraisingly and said, "Well, you're wearing green. She [the spirit] likes that. And you're handsome. So she'll help you. Unless she takes you away. Other *tani* spirits might steal you."

Such a concern is not just a joke or a strange form of a compliment, as the experience of the men fearing traffic accidents indicate. One young day laborer working without a contract at a nearby factory, Lek, who came regularly to the lane, described how he had entered into a relationship with Mother Tani (see also Johnson 2012). He was driving past the mouth of the lane and suffered a severe motorcycle accident. Afterwards, his girlfriend dreamed that the spirit had attempted to claim Lek for her own, and the two had decided that, through no will of his own, Lek had entered into a romantic relationship with the spirit, a relationship that, while dangerous, could nonetheless be turned into something profitable.

The Potency of Accident

The idea of the city that emerges from the devotees of these shrines is a heterodox one. There is a profusion of actants: material, spirit, and human, which together make up an urban ecology beyond capture by any sort of planner. Instead, wild things thought to be safely paved over are found to persist and are harnessed by those also ignored and taken for granted by Bangkok's elite—this is why they are disconnected and even resistant to middle-class plans for a green city. The planner's grid, with its promises of rationality and order, and the galactic polity, with its *mission civilisatrice*, are found to have been tainted from their inception. The city is already undermined. Just as green and yellow dresses hang above black-and-white striped zebras, the presence of nature spirits presents an excess of ownership over seemingly planned space.

Elsewhere, I have examined the uncanny elements of hauntings in (thought-to-be) developed (*jaroen*) space, and how the appearance of ghosts points to anxiety about whether or not the glossy, planned veneer of cities hides something more sinister (Johnson 2013; 2014). But, here, these shrines, while they are places of accident, are also places of potential.

Catherine Malabou, in her *Ontology of the Accident* (2009), turns our attention to the idea of accident as that which transforms the subject into something entirely different and which reveals the self as fundamentally plastic. Malabou focuses on individual transformations: a person is altered by disease or trauma into something unrelated to who he or she once was. Or, for instance, those on the brink of death imagine not how they will be remembered afterwards but realize how they will in fact be forgotten—they see their own transformation into nonperson. While Malabou's remarks strike Western readers as revelatory, such a concept would not be surprising to a Buddhist, where principles of non-self [*anatta*] and impermanence [*anicca*] are, along with suffering [*dukkha*], the Three Marks of Existence [*trailak*]. While many lay people (although, see Cassaniti 2015) ignore the implications of this, observant Buddhists are cautioned to abandon attachment to a persistent self.

The shrines to accident in Bangkok present a surprising parallel. Accidents transform, a transformation that is not a source of fear or anxiety. Jiap seeks not to prevent disaster but rather to harness its destructive power in order to break the karmic trap in which she finds herself: a poor, rural, non-Central Thai-speaking person in the city with neither connections nor good fortune [*kam, chatta*] such as Jiap would have no hope of becoming someone new, but with the transformational power of accident, she can.

Further, accident points to the presence of an author to accident, a thing outside of the human with the power to cause these transformations, a thing already present, but which is revealed at particular moments. Especially in the city, it points out the limitations in the planner's view, and, indeed, the limitations of human agency to control even the most controlled spaces and presents city life as something bound in an ecological network extending beyond the visible. In Bangkok, it is this forest that underlays the city streets that allows for the possibility that its marginalized residents can transcend the urban grid.

Works Cited

Anand, Nikhil. 2017. *Hydraulic City: Water and the Infrastructures of Citizenship in Mumbai*. Durham, NC: Duke University Press.

Ayuttacorn, Arratee, and Jane Ferguson. 2018. "The Sacred Elephant in the Room: Ganesha Cults in Chiang Mai, Thailand." *Anthropology Today* 24 (5): 5–9.

Århem, Kaj. 2016. "Southeast Asian Animism in Context." In *Animism in Southeast Asia*, edited by Kai Århem and Guido Sprenger, 3–30. New York: Routledge.

Baumann, Benjamin. 2014. "From Filth-Ghost to Khmer-Witch: Phi Krasue's Changing Cinematic Construction and its Symbolism." *Horror Studies* 5 (2): 183–96.

Bennett, Jane. 2009. *Vibrant Matter: A Political Ecology of Things*. Durham, NC: Duke University Press.

Cassaniti, Julia. 2015. *Living Buddhism: Mind, Self and Emotion in a Thai Community*. Ithaca, NY: Cornell University Press.

Cruikshank, Julie. 2005. *Do Glaciers Listen? Local Knowledge, Colonial Encounters and Social Imagination*. Vancouver: UBC Press.

DeCerteau, Michel. 1984. *The Practice of Everyday Life*. Translated by Steven Rendall. Berkeley: University of California Press.

Elinoff, Eli. 2017. "Concrete and Corruption: Materialising Power and Politics in the Thai Capital." *City* 21 (5): 587–96.

Evans-Pritchard, Edward Evan. (1937) 1976. *Witchcraft, Oracles and Magic Among the Azande*. Oxford: Oxford University Press.

Ferguson, Jane. 2016. "Terminally Haunted: Aviation Ghosts, Hybrid Buddhist Practices, and Disaster Aversion Strategies amongst Airport Workers in Myanmar and Thailand." *The Asia Pacific Journal of Anthropology* 15 (1): 47–64.

Funahashi, Daena. 2016. "Rule by Good People: Health Governance and the Violence of Moral Authority in Thailand." *Cultural Anthropology* 31 (1): 107–30.

Harms, Erik. 2012. "Beauty as Control in Urban Saigon: Eviction, New Urban Zones, and Atomized Dissent in a Southeast Asian City." *American Ethnologist* 39 (4): 735–50.

Harms, Erik. 2016. *Luxury and Rubble: Civility and Dispossession in the New Saigon*. Berkeley: University of California Press.

Herzfeld, Michael. 2016. *Siege of the Spirits: Community and Polity in Bangkok*. Chicago: University of Chicago Press.

Jackson, Peter. 2010. "Virtual Divinity: A 21st Century Discourse of Thai Royal Influence." In *Saying the Unsayable: Monarchy and Democracy in Thailand*, edited by Soren Ivarsson and Lotte Isager, 29–60. Copenhagen, Denmark: NIAS Press.

Johnson, Andrew Alan. 2012. "Naming Chaos: Accident, Precariousness, and the Spirits of Wildness in Urban Thai Spirit Cults." *American Ethnologist* 394: 766–78.

Johnson, Andrew Alan. 2013. "Progress and Its Ruins: Ghosts, Migrants, and the Uncanny in Thailand." *Cultural Anthropology* 28 (2): 299–319.

Johnson, Andrew Alan. 2014. *Ghosts of the New City: Spirits, Urbanity and the Ruins of Progress in Urban Northern Thailand*. Honolulu: University of Hawai'i Press.

Johnson, Andrew Alan. 2015. "Dreams of the Island Nation: Democracy, Lee Kwan Yew's Legacy, and Modern-Day Mythmaking." *Becoming* (3). March 2015.

Johnson, Andrew Alan. 2016. "Ghost Mothers: Kinship Relations in Thai Spirit Cults." *Social Analysis* 60 (2): 82–96. https://doi.org/10.3167/sa.2016.600206.

Johnson, Andrew Alan. 2020. *Mekong Dreaming: Life and Death Along a Changing River*. Durham, NC: Duke University Press.

Keyes, Charles. 2006. "The Destruction of a Shrine to Brahma in Bangkok and the Fall of Thaksin Shinawatra: The Occult and the Thai Coup in Thailand of September 2006." *Asia Research Institute Working Paper No. 80*. Singapore: Asia Research Institute.

Khamjan, Mala. 2008. *Phi nai Lanna* [Ghosts of Lanna]. Bangkok: Se-ed Books.

Khao Sod. 2009. "*Phi Santika Hian*" [The Angry Ghosts of Santika]. Accessed April 30, 2018. https://hilight.kapook.com/view/35130.

Malabou, Catherine. 2009. *Ontology of the Accident: An Essay in Destructive Plasticity*. Translated by Carolyn Shread. London: Polity Books.

Methanuphab, Chalisa. 2019. "*Baen thung laew pai nai? Chai chiwit to pai yang ngai meua mai mi thung phlasatik*" ["Where is the plastic ban going? How will we live without plastic bags?"]. *Greenery*, December 31, 2019. Accessed October 10, 2020. https://www.greenery.org/articles/insight-ban-plastic-bag/.

M-Thai. 2012. "*Atthan narok santika – thi din thi mi tamnan . . . leuad!*" [Hell-Curse of Santika: Land that Has a Legend . . . of Blood!]. November 2012. Accessed April 30, 2018. https://shock.mthai.com/story-shock/1473.html.

Pearson, Trais. 2020. *Sovereign Necropolis: The Politics of Death in Semi-Colonial Siam*. Ithaca, NY: Cornell University Press.

Peleggi, Maruizio. 2002. *Lords of Things: The Fashioning of the Siamese Monarchy's Modern Image*. Honolulu: University of Hawai'i Press.

Povinelli, Elizabeth. 2016. *Geontologies: A Requiem to Late Liberalism*. Durham, NC: Duke University Press.

Rademacher, Anne. 2011. *Reigning the River: Urban Ecologies and Political Transformation in Kathmandu*. Durham, NC: Duke University Press.

Raffles, Hugh. 2002. *In Amazonia: A Natural History*. Princeton, NJ: Princeton University Press.

Schivelbusch, Wolfgang. 1988. *Disenchanted Night: The Industrialization of Light in the Nineteenth Century*. Translated by Angela Davies. Berkeley: University of California Press.

Siegel, James. 2006. *Naming the Witch*. Stanford: Stanford University Press.

Swearer, Donald and Sommai Premchit. 1998. *The Legend of Queen Cama: Bodhiramsi's Camadevivamsa, a Translation and Commentary*. Buffalo, NY: SUNY Press.

Tambiah, Stanley J. 1970. *Buddhism and the Spirit Cults in North-East Thailand*. Cambridge: Cambridge University Press.

Textor, Robert B. 1973. *Roster of the Gods: An Ethnography of the Supernatural in a Thai Village*. New Haven, CT: HRAF.

Winichakul, Thongchai. 2000. "The Others Within: Travel and Ethno-Spatial Differentiation of Siamese Subjects 1885–1910." In *Civility and Savagery: Social Identity in Tai States*, edited by Andrew Turton, 38–62. Surrey: Curzon Press.

2
Putting the Garden Back: Cultivating Life through Urban Gardening in India[1]

Camille Frazier

In May 2016, an environmental scientist at the Indian Institute of Science, T. V. Ramachandra, made headlines with his assertion that Bengaluru (formerly Bangalore)[2]—the heart of India's information technology boom and one of its fastest growing cities—would be uninhabitable and "dead" in five years (Menezes 2016). The claim struck a nerve and led to a series of English-medium news articles and social media posts debating Ramachandra's statement (Deviah 2016; Srinivasan 2016). As evidence of its continuing resonance, the discussion again surfaced in August 2017, with headlines proclaiming Bengaluru's demise in three years (Thakur 2017). Although Ramachandra's contentious argument was based on his study about environmental degradation in the city (Bhat et al. 2015), he expressed the effects of this decay in terms of food and health. He is quoted in the *Deccan Herald* as saying, "what's the point [of] earning better when the food that you eat is adulterated? As a result of unplanned urbanisation, Bengaluru is going to be an unliveable and dead city in the next five years" (Menezes 2016). With this bleak prediction, Ramachandra challenged narratives proclaiming the economic advantages of rapid urban development—higher wages and a burgeoning middle class—by suggesting that money means little when one's food is inedible and the urban environment is uninhabitable.

Ramachandra's claim captures a popular narrative about Bengaluru and reflects pervasive concerns about the role of shifting economies in ecological decay and the un-livability of developing cities. Such concerns are widespread in Asia, as is well documented in this volume. Often, this unease is expressed through narratives of food and health. Whether about the dangerous effects of adulterated milk or artificial ripening, anxieties about eating pervade life in Asia's cities (Solomon 2015; Tracy 2010; Yan 2012). These fears are linked with the growing concern that unscrupulous actors in the food supply chain are wittingly duping urban consumers

---

1. Portions of this text were previously published in 2018. "'Grow What You Eat, Eat What You Grow': Urban Agriculture as Middle Class Intervention in India," *Journal of Political Ecology* 25 (1). https://doi.org/10.2458/v25i1.22970. I thank the *Journal of Political Ecology* for permitting the reuse of the material in the chapter.
2. In November 2014, the city's official name was changed from Bangalore to Bengaluru.

who are largely alienated from food sources and therefore unable to "know" what they are eating. This sense of estrangement leads to projects such as those discussed by Caroline Merrifield[3] in this volume, in which regional foodsheds become sites for remembering the past, contesting the present, and reconfiguring the future. Such fears and the projects that they motivate are not unique to Asia, of course, but they capture the sense that the rapidly developing cityscapes that characterize many parts of Asia today are producing new urban ecologies and communities.

In this chapter, I analyze organic terrace gardening in Bengaluru, a burgeoning movement that engages directly with the life and death of urban ecologies. An increasing number of middle-class residents of the city are growing fruits and vegetables for home consumption, motivated by two primary concerns: first, worsening food safety and health conditions, and second, declining green spaces in the city. These concerns are best understood within a context of rapid urban development that has altered how middle-class residents perceive their food and urban ecologies and the effects of these ecologies on their health.

As India's "IT capital," Bengaluru is one of India's fastest-growing cities. With an urban population that tops 9.5 million and a decadal growth rate of 46.68 percent between 2001 and 2011, the city has changed drastically in recent decades (Government of India 2011). In this context of urban upheaval, efforts to "grow what you eat, eat what you grow" represent an attempt among the urban middle class to intervene in decaying urban bodies and ecologies (Frazier 2018). By cultivating organic food for home consumption, organic terrace gardeners create life in two forms: healthy human bodies and garden spaces. They focus on producing and disseminating the skills of cultivation as a way to "put the garden back in the Garden City," in the words of one organic terrace gardening advocate (personal communication). Gardening thus offers an opportunity to cultivate "life" in the midst of impending "death," and offers a site to intervene at multiple scales, from the individual body to the expanding cityscape. However, in focusing on particular methods and sites of cultivation—namely, organic gardening in small, private spaces such as rooftops or apartment balconies—organic terrace gardeners' practices remain specific to the middle class and neglect other historical and contemporary forms of urban food production.

In what follows, I present data from eighteen months (June 2014 through January 2016) of participant observation at urban gardening fairs and workshops, and interviews with organic terrace gardeners, most of whom self-describe as "OTGians," a title associated with the Organic Terrace Gardening (OTG) Facebook group. What visions of life and death in the city motivate middle-class individuals to take up organic terrace gardening? How do organic terrace gardeners' efforts compare with other forms of urban food production? What do these junctures and tensions teach us about the possible futures of urban agriculture as cultivating life

---

3. See Caroline Merrifield, "Keeping Pace with the Foodshed in Hangzhou," chapter 9 in this volume.

in the face of urban decay? OTGians use gardening to intervene in the changing food networks and urban ecologies that they understand to have negative effects on themselves, their families, and their city. In so doing, they have created a vibrant community dedicated to sharing resources and knowledge about urban gardening. However, because OTGians' efforts are rooted in individual, class-specific experiences of the transforming cityscape, the OTG community remains limited to the middle class. Its vision for urban food production is rooted in historical and contemporary forms of urban development that exclude other food producers and practices from the city.

Cultivating the City

In producing food for household consumption, OTGians are engaged in a form of "urban agriculture," although they rarely use the term to describe their efforts. Existing analyses of urban agriculture often focus on the use of vacant spaces in cities as the primary site of urban food production, either in the form of community gardens in the global North (Chung et al. 2005, Hite et al. 2017; Poulsen et al. 2014; Sokolovsky 2011) or as a livelihood strategy in the global South (Drakakis-Smith, Bowyer-Bower, and Tevera 1995; Bryld 2003; Simatele, Malula, and Binns 2008). In contrast, OTGians garden in individual private spaces, yet their motivations reflect global forms of activism around food, including distrust of the global food system, concerns about production practices and their effects on human and environmental health, and desire to bring food production and green spaces to the city.

Like many alternative food movements around the world, OTGians are embedded in broader structures of inequality that limit participation to those who have the time, space, and resources to access healthy and sustainable food (Guthman 2008; Pudup 2008; Slocum and Cadieux 2015). The class specificities of the OTG community are made immediately apparent by the phrase used to describe their efforts: "organic terrace gardening." This phrase marks practitioners as middle class for two reasons: first, it is in English (I will return to this point later in this chapter); and second, it requires access to a private space, whether the rooftop of an individual house or an apartment balcony, where plants can be grown for household consumption. In using language and private space to describe OTGians as middle class, I employ the class schema outlined by Fernandes and Heller (2006) that considers linguistic, caste, and educational inequalities as inseparable from property and income-based class hierarchies. Although the term "middle class" captures a wide breadth of lives and livelihoods, I find it useful in highlighting the educational, linguistic, professional, and spatial forms of class distinction that characterize and are reproduced through the OTG community.

Figure 2.1: A rooftop terrace garden in Bengaluru. Source: author, 2015.

OTGians' garden spaces vary widely and run the gamut from a few pots on a balcony to an entire terrace full of plant beds.[4] These spaces of cultivation offer a few key insights into the activities of OTGians. For one, OTGians focus their efforts on what Sidney Mintz calls "fringe" foods, which help to "enliven" and "enhance" the complex carbohydrates that make up the core of agrarian foodways (Mintz and Schlettwein-Gsell 2001, 41). Largely due to the restrictions of gardening on one's balcony or rooftop, OTGians grow fruits, vegetables, and herbs, rather than the grains and legumes that make up the bulk of the Indian diet. However, OTGians also justify their focus on "fringe" foods by suggesting that in Bengaluru today, it is much easier to purchase certified organic grains and legumes than fruits and vegetables. Whether certification can be trusted is another issue altogether (and one that will return later in this chapter), but many of the OTGians with whom I spoke said that they were fairly confident in the quality of certified organic grains and legumes but were very concerned about the fruits and vegetables available in the marketplace. This justification for self-provisioning, and the purchasing power that it implies, highlights a second key point: that OTGians are mostly middle-class

---

4. In Bengaluru, the English word "terrace" generally refers to the flat cement rooftop of a house, but the phrase "terrace gardening" now captures a larger range of private spaces used for food production, including balconies and small yards.

individuals whose lives and livelihoods are rooted in processes of urban development that have reconfigured the urban ecology. The life that the OTG community generates on apartment balconies and rooftops is rooted in the historical and contemporary displacement of market-oriented gardens and the exclusion of a caste-specific gardening community.

Shortly after my arrival in Bengaluru, I started to hear about the Vahnikula Kshatriya (also known as the Thigala) caste, renowned for its horticultural prowess. The Vahnikula Kshatriya (VK) community is responsible for the Karaga festival, one of the city's largest and most famous religious events. The caste falls under the Other Backward Classes (OBC) category established by the Indian government, which includes the lower (but not the lowest) castes that have been disadvantaged and today receive a certain percentage of reserved positions in public sector employment and education.[5] Historically, VK community members owned farmland near the city's manmade lakes and supplied much of the food sold to urban consumers (Srinivas 2001). For many years, they were the primary horticultural producers for the city, and to this day they are known for their gardening and landscaping skills. Over time, as the priorities of urban development changed, the lands where VK families cultivated food were put to uses that largely excluded the community altogether.

Chennappa, a retired government bureaucrat and leader in the VK community, presides over a credit association office located on a narrow street in the city's administrative center. Seated at his desk at the end of a long boardroom table, Chennappa explained to me that for centuries, his community used to grow fruits and vegetables for the Bengaluru market. Now, most of the community's land, especially nearby the city center, has been usurped by the city government for "development." He estimated that in the past, "eighty percent of Bengaluru land was cultivating and growing vegetable and fruits." This started to change in 1933 under British rule, when 210 acres of VK lands were "acquired" for Cubbon Park, 110 acres for a housing colony, and 88 acres for developing a road. This began the process of displacement, and in 1938 things changed for the worse: "earlier to that [1938], on request they [city government authorities] used to take and develop the layouts," but later, "against the will of the agriculturalist they started acquiring." The trend continued after Indian independence in 1947, and in 1951 the City Improvement Trust Board (CITB) was founded (it is now the Bengaluru Development Authority): "without giving any importance to the gardening and production of vegetables for the city, they [the CITB] acquired land. See, this whole area [was] growing fruits and vegetables. They acquired and closed it," Chennappa explained as he swept his hand around us, drawing in the building and surrounding area. His own family experienced this loss firsthand—their lands were acquired for urban development projects when he was a child and he was forced to find wage labor.

---

5. The most marginal caste groups, the Dalit and tribal communities, are excluded from the OBC and instead belong to the Scheduled Castes and Scheduled Tribes categories, respectively.

It is in these converted spaces that many OTGians now reside, in well-established apartment complexes or densely packed single-family homes. Many are professionals working in the information technology (IT) and affiliated industries, often in higher-paying positions as managers, developers, and upper-level technicians. As such, they embody the aspirations and insecurities of the burgeoning middle class, from the clothes they wear to the apartment buildings they call home. While they recognize the benefits of working in the IT industry, they express a sense of concern about the longevity of their careers and lifestyles. As one OTGian told me, the global economic downturn in 2008 convinced him that his career as an upper-level programmer in a multinational firm is less secure than being a farmer, since it does not ensure access to life's basic necessity: food. Such feelings of insecurity reflect class-specific experiences of Bengaluru's shifting cityscape and are key to understanding the concerns that motivate the "grow what you eat, eat what you grow" philosophy and practice.

Organic Terrace Gardening as a Middle-Class Project

"Start any idea from your house and your neighbors will follow," explained Anand, a founding member of one of the largest and best-known terrace gardening associations in India. During my fieldwork with the organization, attending fairs and workshops, I heard the story of his personal transformation several times: he was a scientist at an agricultural university, working on the propagation and uptake of Green Revolution technologies to manage pests. With time, he began to question the effects of these chemically intensive methods for pest control. He quit his position and committed himself to spreading the word about the harmful effects of Green Revolution technologies and teaching alternative ways of food production.

Several years ago, when his airplane was forced to circle above Bengaluru before landing, he noticed the bare rooftops littering the cityscape below. This experience gave him the idea to promote terrace gardening as both a way to decrease the consumption of pesticide-laden fruits and vegetables and to add green spaces to the city. In 2005, he and a small group of urban professionals began conducting workshops, and in 2011 they created a trust focused on promoting organic terrace gardening in Bengaluru. Since then, the OTG community has grown into an extensive network. Much of the action is online, where the Organic Terrace Gardening Facebook group provides a space for OTGians to share successes and failures, ask questions and provide answers. At the time of writing, the OTG Facebook group had over 33,000 members and was full of photos from avid gardeners sharing images of their harvest or asking for answers about a particular pest or problem.

In addition to its online presence, the trust puts on quarterly fairs that rotate to different parts of Bengaluru called Oota From Your Thota ("food from your garden"). These events and their associated workshops create an energizing sociality and sense of community built around the knowledges and practices of organic

Figure 2.2: An Oota From Your Thota event. Source: author, 2014.

terrace gardening. The fairs are popular and well attended, and the organizers estimate that they have around five thousand visitors during each single-day event. The organizers generally accept sixty vendors per fair, and there is always a waiting list. Vendors are mostly newer companies that were established to meet urban interest in gardening and offer products like self-watering pots and composting bins. The Oota From Your Thota fairs are meant to aggregate in one place everything necessary to start an organic terrace garden, and they are successful in this regard—visitors can find everything from seeds to soil to pots. The goal, as the organizers explained it, is to promote a "holistic transition" to an organic lifestyle centered on the motto, "grow what you eat, eat what you grow."

As with the phrase "organic terrace gardening," language is one of the key ways in which these fairs are marked as middle-class events—workshops, handouts, and vendors' signs are in English. This matters because language is a site of conflict that represents larger battles over the future of the expanding city (Nair 2000). It reflects educational and professional exclusions that mark class differences and concern about the influx of outsiders (of different class positions) into Bengaluru. From its name alone, the Oota From Your Thota fairs are implicated in these shifts, both as evidence of them as well as counterstrategies for maintaining the city that came before. In a February 2016 news article in *The Hindu* titled "Bengaluru's Growing

Pride," the author labels Oota From Your Thota "a perfect phrase" because it captures how the "Garden city is grafted with IT city, to create the new-age urban farmer who harnesses technology and knowledge to grow a green organic spread" (K. 2016). In such descriptions of the OTG community, the city's burgeoning class of IT professionals is explicitly linked with particular histories and futures of the city.

The interest in "a green organic spread" among the "new-age urban farmer" is rooted in two primary concerns: first, fear of the health effects of unsafe food and untrustworthy food producers; and second, concern about the loss of green spaces in the city. I consider these in turn.

Fears about Food Safety

The detrimental health effects of pesticide residues are a common topic in Bengaluru today and appear often in news media as well as day-to-day conversation. There is a range of ways in which urban consumers attempt to manage these dangers, and organic terrace gardening has become one of the most visible. Pesticide contamination was one the first topics addressed in every terrace gardening workshop I attended, whether conducted by Anand's trust, the Karnataka Department of Horticulture, or companies selling ready-to-grow kits for beginners. The same image appeared often: a baby crying with his eyes closed and fists balled, his horrendously large head occupying the majority of the photograph. As the presenters explained, this image captures the deformities experienced in the state of Kerala in recent generations as a direct result of aerial spraying of the pesticide endosulfan.[6] Narratives of endosulfan poisoning are common among activist circles working against the effects of the Green Revolution, but concerns about pesticide residues extend much further than the Kerala tragedy. A 2015 graphic in *The Times of India*, for example, lists seven "common banned pesticides used" and states that "short term exposure" to pesticides can cause a range of negative health effects, from a "headache" to "slurred speech" (*Times of India* 2015).

Concern over pesticide residues and the health consequences of conventional agriculture appeared frequently in my interactions with organic terrace gardeners. One woman at a terrace gardening workshop explained that her goal in attending the event was to learn how to start a garden at her parents' house since her parents have "some or the other ailment or some kind of disease" and she was confident that it was because "the water and the food they eat was full of pesticides." The workshop organizer responded that this kind of worry is what inspired their workshops. Growing your own food, to the extent possible in the space that you have, is the only way to limit the harmful effects of chemical-laden foods, he suggested.

---

6. In January 2018, the Indian Supreme Court directed the state of Kerala to pay INR 5 billion in compensation to victims of endosulfan poisoning.

Due to the perceived dangers of pesticides and other industrially produced inputs like urea (these are often conflated under the term "chemicals"), an organic method of production is central to how terrace gardening is practiced in Bengaluru today. The term "terrace gardening" is almost always assumed to be "organic." This is not only true in the events put on by Anand's trust, which is explicitly against the technologies of the Green Revolution, but also in government programs and corporate product lines aimed at urban terrace gardeners. In a Department of Horticulture organic terrace gardening workshop, the agricultural scientist who led a presentation about cultivation methods began with a slide titled "poisonous vegetables." He was quick to point out that the organic methods discussed in the workshop were intended only for urban terrace gardening, because organic production is ineffective for "real farmers" who are worried about yield. So, he suggested, the only way to limit your chemical intake is to grow your own fruits, vegetables, and medicinal plants.

A suite of fears and ethical commitments often accompany organic discourses and practices. Chemical-intensive agriculture, genetically modified organisms (GMOs), non-native and hybrid seeds, and processed foods are often conflated as equally pernicious. During a seed exchange at a public park in the city center, two passersby interrupted the group's activities to advertise the International Yoga Day festivities that were scheduled in the park for the following day. Handing out fliers, the young men explained they were with the BJP (the political party of the current Prime Minister, Narendra Modi, who was credited with—and critiqued for—establishing International Yoga Day). They said that while walking through the park, they overheard our meeting and were very happy to see such activism around the issue of organic food, which they support. In response, a vocal middle-aged woman pointed her finger at the young men, saying, "you tell [Prime Minister] Modi that we don't want any GM [genetic modification]." One of the young men replied that they agree with her, but it is a very complicated issue. The woman retorted, "if GM, no PM!" Everyone in the seed exchange group laughed and clapped. The two BJP spokesmen smiled uncomfortably and quickly finished handing out their fliers. I was surprised by the interaction, because I had not yet realized the importance of the GM issue to members of the organic terrace gardening group. With time, I came to see this event as one of many that emphasized organic terrace gardeners' commitments to a range of issues that they saw as interconnected, primarily those surrounding Green Revolution and New Green Revolution technologies and methods.

It is not only these interlinked concerns about GM and chemical overuse that worry organic terrace gardeners but also their inability to trust food producers. Key to the OTG narrative is the belief that agricultural practice in general, and the organic market in particular, is highly suspect. Such fears are explicitly linked to the work of unscrupulous actors who are careless about their practices' impacts on consumers' health. This reflects broader concerns about what Harris Solomon (2015) calls the "reliability" of food and eating in urban India. Even while the organic and natural foods market expands in India and elsewhere, there remain many questions about its efficacy and impact (Aistara 2018; Galvin 2018; Guthman 2004). This means that while many OTGians shop at the growing number of organic retailers around the city, many are skeptical that these foods abide by organic standards. Through my research with organic food retailers, I learned that this is a commonplace concern that retailers must address in their relationships with customers. Often, customers go to great lengths to decide whether they can trust a particular brand or retailer—for example, by calling food companies to ask questions about their production practices, or building a personal relationship with a storeowner.

Yet for many OTGians, these tactics for establishing trust do not go far enough. As we stood around having lunch during a day-long terrace gardening workshop, I struck up a conversation with a man who seemed particularly enthusiastic about the day's activities. In answering my question about what motivated him to attend the workshop, he explained that he began gardening about three months ago when his wife started buying organic products. He argued with her that if they were going to be paying more for their food, he wanted to know whether he could trust that the organic products are "really organic." Even if he took the time to visit the organic farms, "what knowledge do I have to find out if it is an organic food or not?" So, he decided to instead grow his own organic vegetables, as this was the only way he could have full confidence that what he consumed was truly organic.

As this example illustrates, distrust of organic certification pervades OTG discourse. As gardeners asked me on several occasions, how can you be sure that food labeled organic is *actually* organic? Even if a product is certified, there is no guarantee. More effective than certification is "trust" and "belief" in a particular farmer or group, they suggested. But because this does not come easily to urban residents, who find it hard to build direct relationships with food producers, the best course of action is to grow what food you can so that you can be "assured" of its safety.

Cultivating Nature

While concerns about food safety and untrustworthy food sources were the primary motivation for the majority of OTGians, many also emphasized the role of their terrace gardens in creating nature spaces in the city. In emphasizing the role of food cultivation in creating nature spaces, OTGians create a different sort of urban ecology than those of the "luxury" and "eco-friendly" apartments and gated complexes that boast manicured lawns and lakes as recreational spaces. Properties by the ZED group, for example—with names such as ZED Woods and ZED Earth—advertise both self-managed waste systems and swimming pools. The OTGians with whom I interacted did not live in these complexes—most often, my interactions with OTGians led me to homes in older neighborhoods in the heart of the city or in its more established suburbs, rather than newly constructed gated complexes and high-rise apartment buildings in the urban fringe.[7] In part, OTGians' residence in more established neighborhoods of the city made them all the more committed to "putting the garden back" in the areas of the city for which Bengaluru has earned its nickname of "Garden City," a title that captures the role of horticultural fields, boulevard trees, and bungalow and botanical gardens in the city's imaginary (Nagendra 2016; Nair 2005; Srinivas 2001).

Among the very first slides in Anand's introduction to his terrace gardening workshop was a picture of traffic in Bengaluru. He paused on the image, saying that the city has become "too big." He suggested that with the "IT invasion of Bangalore" a lot has changed—thirty-five to forty years ago every house had an "ornamental garden in the front and a kitchen garden in the back." Now, the city has lost its greenery, he said. Every organic terrace gardening workshop that I attended and almost every OTGian with whom I spoke referred to how dense the cityscape had become and how much of its gardens and green spaces had been lost. For many, the organic terrace garden offered a space to re-create past memories and idealized nature spaces, thereby countering the trend toward a cityscape that is better known for its traffic than its trees and gardens.

Srinath's passion for building an "urban jungle" was immediately visible from the street—his house, located in an upper-middle-class and upper-caste neighborhood in central Bengaluru, was completely covered with lush green vines with big, purple flowers. During my tour of his garden, Srinath explained that his neighbors get angry with him for planting along the curb because it reduces parking space in an already tightly packed neighborhood with large, individual homes.[8] Yet he does not mind his neighbors' frustrations, he said, because he is gardening to "go against the norm." Srinath was "against the norm" in more ways than one—the son of a poor farmer, he had struggled to attend school as a child but managed to make a successful career as a business consultant in Bengaluru. As we walked through his rooftop garden, Srinath pointed out that he was creative in using every kind of object he could find to grow plants—sprinkled among the terracotta pots were old cement bags, a bathtub full of water plants, and even a Western-style toilet. Speaking with pride about the diversity of species on his rooftop, Srinath explained that building a "whole ecosystem" is his "passion." He loves the birds, insects, monkeys, and stray dogs and cats that come to enjoy his "jungle."

Building "ecosystems" and creating connections with "nature," especially for urban children who are believed to otherwise lack access to such experiences, are primary motivations for organic terrace gardeners. Both men and women emphasize

---

7. This position in the more established neighborhoods of the city might mean that OTGians are less likely to experience feelings of exclusion and displacement that can create unlikely solidarities between lower-middle-class residents of peripheral housing developments and the working-class poor, as discussed by Ranganathan (2011).
8. As Srinath's experiences indicate, OTGians do not always have the easiest relationships with their neighbors. This contradicts Anand's assertion that by growing their own food, OTGians will encourage their neighbors to take up gardening as well.

Figure 2.3: Srinath's "urban jungle." Source: author, 2015.

the joy in bringing their children into their gardens, showing them "where food comes from." For middle-class children growing up in apartment complexes and tightly packed urban layouts, access to this knowledge is limited (Nagendra 2016). Their parents understand this alienation from nature and food sources as damaging to children's development, and organic terrace gardening offers one way for urban middle-class families to fill these gaps in their children's education. Shruthi, for example, was committed to growing a terrace garden so that her children could spend time playing with soil and being in natural spaces. She valued her childhood memories of playing in her mother's garden in what was then a sleepy neighborhood in central Bengaluru, and she was worried that her children would not have access to the same experiences while growing up in their densely packed upper-middle-class neighborhood. As a native of Bengaluru who was also employed in the IT industry, she felt that gardening was the best way to maintain the city that she remembered and loved, even while she participated in the industry that she held responsible for the changing cityscape. After a long commute through the city's increasingly unbearable traffic, she said, the garden was the only place she could relax.

As Shruthi's and Srinath's stories suggest, the effect of cultivation is critical to understanding what OTGians produce in the terrace garden. The embodied experiences of mixing soil, planting seeds, and tending plants are often as valuable as the resulting foods themselves. Many in the OTG community devoted a large percentage of their leisure time to working their gardens or attending affiliated events in the city, such as seed exchanges.[9] While the far majority of these individuals employed domestic servants to take care of a variety of household tasks, they were themselves responsible for maintaining their gardens, as well as many of the associated tasks, such as composting. This was often explained to me in two ways: first, that experiencing the work of cultivation was the point of having an organic terrace garden, and second, that they did not trust that their workers would have the knowledge and skill necessary to care for their plants. I will consider this second point in more detail below, but here I wish to emphasize that both the affective practices of cultivation and the associated knowledges produced by OTGians are considered of primary value. Often, these practices and knowledges are made all the more valuable by their connection with personal and historical memories of gardening in Bengaluru.

Class and Caste in the Garden

In focusing on gardening as a way to "grow what you eat, eat what you grow," OTGians offer a critique of unbridled urbanization and emphasize the importance of creating and maintaining green spaces not only for leisure but also for food production. This challenges a singular vision for the Garden City that focuses on urban nature spaces as sites of recreation rather than livelihood (Unnikrishnan and Nagendra 2014). However, their approach to urban food production reflects historical changes in the meaning of the word *garden* that privileged individual household gardens over market-oriented fields. Smriti Srinivas (2001) charts a change in approaches to gardening in Bengaluru between the precolonial and colonial periods. Specifically, gardens in the precolonial period "were essentially horticultural lands where a variety of fruits, flowers, and vegetables were produced for the urban centers," while "gardens built by the British were large or small grassy parks with trees and flowers and surrounded by suburban bungalow houses with their own green spaces" (Srinivas 2001, 47–48). By lamenting the loss of bungalow gardens in their descriptions of urban transformation, OTGians' narratives point to a very particular version of the Garden City, one that reflects the colonial and

---

9. This emphasis on leisure time also meant that gardening created conflicts related to gender and domestic labor—Shruthi, for example, was critiqued by her husband and parents-in-law for favoring her garden over her family.

postcolonial emphasis on private spaces and middle-class experiences of urban transformation (Nair 2005).

Those who produced food commercially for the city have a different history of the Garden City and a different experience of decayed garden spaces. As described by Chennappa earlier in this chapter, the VK community's experience of urban transformation is anchored in disenfranchisement and exclusion. This story of urban transformation does share some similarities with those of OTGians. At the most basic level, they are narratives of loss—of land captured by the expanding city, and unhealthy bodies and ecologies left behind. But the land, labor, and futures caught up in these stories diverge. Narratives of bungalows with decorative gardens in the front and kitchen gardens in the back conjure a particular experience of gardening in private spaces in Bengaluru, one rooted in the very urban development that has displaced horticultural fields in favor of individual homes and apartment complexes. In their descriptions of the Garden City's past, OTGians often reference a specific history of the city while remaining unspecific about when, by, and for whom the Garden City earned its name. By erasing the caste and class specificities of how the city was and is gardened, the OTG narrative represents middle-class experiences of Bengaluru's decay as universal and positions the greening of private spaces as the most prominent roadmap for the future. This further marginalizes the histories and futures of particular caste communities.

The founders of the organic terrace gardening trust are aware of the class inequalities in their efforts. I was often impressed by their self-critical reflections on their inability to challenge existing hierarchies. Anand admitted to me early on that their primary audience is middle-class families. He explained that the urban elite do not participate because "they leave everything to their servants," while the lower class is "too difficult to reach." Attempts to account for some of these class inequalities appeared in different ways during my interactions with the organization and its founders. For example, I learned that the trust had attempted to establish a gardening initiative targeting slum communities in Bengaluru. The idea was to introduce climbing vegetables, which can be trained up onto the roof of the small and densely packed homes where they have sun and space to grow. However, Anand lamented that the initiative never came to fruition because they were unable to secure funding for the project.

Despite such attempts at crossing the class divide, the ideologies and practices of urban food production among the OTG community produce a different, and in some ways contradictory, vision for the Garden City than that of the VK community. For one, the knowledges, practices, and materials specific to organic terrace gardeners—from finding organic compost to watering potted plants—create different practices, affects, and ethics of cultivation than those of market-focused producers. This makes the organic terrace garden a site of class- and caste-based distinction, both in OTGians' relation to market gardeners and also in relation to their domestic servants, whom they often do not entrust with gardening tasks. In addition, by emphasizing organic methods as a way to limit the harmful effects of chemically intensive agriculture, organic terrace gardeners distinguish themselves from urban farmers who produce for market and often rely on chemical fertilizers, pesticides, and water from sewage drains and contaminated lakes. This means that although market-oriented urban gardeners also contribute to creating green spaces in the city and cut down on food miles, they do not fit into the OTG community's understanding of healthy bodies and ecologies. Rather, they are a source of OTGians' anxieties—urban farmers who rely on chemical inputs and contaminated water sources to grow fruits and vegetables for the market are the very producers mistrusted by organic terrace gardeners.

Another point of departure between OTGians and VK gardeners is how these individuals understand their efforts in relation to broader structures and processes. OTGians have created a vibrant community of gardeners, but it is one predicated on individual rather than structural interventions. I do not mean to suggest that gardening in private rather than public spaces forecloses civic engagement (Chung et al. 2005). Rather, I wish to highlight that OTGians place responsibility for safe foods and healthy urban ecologies on individual households. It is up to individual OTGians to "put the garden back into the Garden City." This is the entire point—as Anand suggests: the goal is for each family to establish a home garden. Still, OTGians understand their interventions to have an impact beyond themselves—in addition to inspiring others to take up gardening, individual gardens can mediate, at least to some extent, the forms of urban degradation and un-livability that most concern organic terrace gardeners. It is critical to note, however, that OTGians' individualized efforts might preclude activism toward more structural interventions into Bengaluru's shifting food ecologies. As with Erik Harms' discussion of floods in Saigon (this volume),[10] there is the possibility that such individualized efforts will obscure broader patterns and thereby deepen rather than alleviate the insecurities that OTGians wish to address.

Chennappa's description of the reasons why his caste community members continue to farm in small urban pockets and the city's outskirts—despite the community's systematic exclusion and removal from the developing city—illuminates a different understanding of the relationship between gardening and urban life:

> We feel that this is our family culture. We have to maintain it. . . . Just imagine, one day vegetables [will] not [be in the] market. What are you going to do? We feel, and we satisfy [ourselves], we are producing something for the society. . . . See, very healthy and good vegetables [VK farmers are] growing and giving [selling] to society. And good fruits they're growing to give society. And green vegetables. Flowers! To ladies, women, and also to the temple. They are giving. But they do not have a piece of flower for their own [hair]. See, that is the condition. They do not have two *saris* to change, but they are giving healthy food to the society.

---

10. See Erik Harms, "Concrete Ecology: Covering and Discovering Saigon's Ecology in a Time of Floods," chapter 8 in this volume.

For Chennappa, VK contributions to society means that urban food production among the VK community should be valued and promoted, which makes the economic inequalities that keep the community from enjoying the fruits of their labor especially frustrating. But he recognizes that certain kinds of knowledge and labor are more valued than others. In Chennappa's words, "nowadays it has become very difficult to lead a normal life. Because the software [industry] has come, everybody is educated, and our caste people are not well educated. Only middle-class and rich peoples' sons, children are well educated." The solution to this problem, for Chennappa, will have to come from both the community—he encourages his fellow VK parents to send their children to school, a point of tension in the community— as well as from government intervention. He advocates, then, for a structural solution to what he understands to be a structural problem. As he put it, "lip sympathy will not work."

Conclusion

OTGians' efforts to intervene in Bengaluru's shifting food ecologies differ from Chennappa's both in form (organic versus conventional agriculture) and in scale (the size of the garden and its output, as well as the attention to individual versus structural solutions). At a more fundamental level, the processes of urban development that have given life to one form of gardening and displaced another mean that these communities have radically different relationships with the shifting urban ecologies. However, it is worth asking whether these different practices and priorities might be brought together under a shared ethic of cultivation. In refocusing attention on food production, and especially in emphasizing teaching and knowledge sharing around food production as a skill, OTGians call into question the priorities of urban development. They use gardening to create alternative spaces and practices of community building, and in so doing intervene in the food systems and urban ecologies about which they are concerned. At the same time, however, organic terrace gardeners are members of the educated, English-speaking middle class who see their work in the terrace garden as an alternative to their professional work in the very industries that have displaced other cultivators. The OTG community's intervention rests on the class-based inequalities in education, language, and types of work that have marginalized lower castes and classes from Bengaluru's past, present, and future. They are in this way similar to other environmental interventions among India's urban middle class that often strengthen structural inequalities in the name of urban "greening" (Baviskar 2002). This interlinking of environmental activism and class exclusion appears as a pattern in much of Asia, as discussed by scholars in this and other volumes (Rademacher and Sivaramakrishnan 2013; 2017).

My goal in making this critique is to highlight the potential for solidarity across scales, practices, and sites of urban agriculture as a life-building process. As Anne

Rademacher notes in her review of the literature on urban political ecology, there is a "constellation of competing and meaningful understandings of urban nature, each potentially located in a privileged or empowered social position at different moments" (2015, 138). She suggests that these competing understandings can be "generative of new affinities, sometimes surprising political maneuvers, and distinctly moral social logics" (2015, 142). There is space for such new affinities among food producers in Bengaluru. For example, Rajinappa, a young VK man, told me that he would continue to cultivate his family's small plot of land next to a congested highway overpass because farming is still valued by his community and it is important for him to maintain green spaces in his city. Rajinappa's dedication to cultivating a verdant urban ecology is similar to that of many OTGians, and I find it energizing to imagine Rajinappa's field and Srinath's "urban jungle" as part of a shared pursuit. Might the interplay among these perspectives offer space to cultivate new formations of urban life and livelihood that address ongoing exclusion and move toward a different future?

I do not imagine that simply bringing together the OTG and VK communities will solve decades-long processes of displacement, but there is space to cultivate a different life of nature in Bengaluru. OTGians' commitments to healthier futures for themselves, their communities, and their urban ecologies are impressive and

Figure 2.4: Rajinappa's garden, located next to a major highway overpass. Source: author, 2015.

potentially transformative. By reevaluating and reworking the relationship between food and urban ecologies, OTGians can generate new approaches to urban food security and sustainability. They have captured the concerns and experiences of the urban middle class to build a community committed to gardening as an answer to worsening food safety and declining green spaces in the city. However, the class and caste distinctions that divide organic terrace gardening from other forms of urban food production are anchored in processes of urban development that privilege certain lives, livelihoods, and futures above others. There is room to expand the meanings and practices of urban gardening in Bengaluru, for the benefit of the city and its diverse communities.

Works Cited

Aistara, Guntra A. 2018. *Organic Sovereignties: Struggles over Farming in an Age of Free Trade.* Seattle: University of Washington Press.

Baviskar, Amita. 2002. "The Politics of the City." *Seminar: A Symposium of the Changing Contours of Indian Environmentalism*, no. 516.

Bhat, V., B. H. Aithal, and T. V. Ramachandra. 2015. "Spatial Patterns of Urban Growth with Globalisation in India's Silicon Valley." Conference Paper. Proceedings of National Conference on Open Source GIS: Opportunities and Challenges. Varanasi, India. October 9–10, 2015.

Bryld, Erik. 2003. "Potentials, Problems, and Policy Implications for Urban Agriculture in Developing Countries." *Agriculture and Human Values* 20 (1): 79–86. https://doi.org/:10.1023/A:1022464607153.

Chung, Kimberly, Robert J. Kirkby, Chet Kendell, and Jo Ann Beckwith. 2005. "Civic Agriculture." *Culture & Agriculture* 27 (2): 99–108. https://doi.org/:10.1525/cag.2005.27.2.99.

Deviah, M. A. 2016. "Bengaluru Won't Die in Five Years; It's Already Dead. *Firstpost*, May 5, 2016. Accessed October 1, 2020. https://www.firstpost.com/living/bengaluru-concretisation-urbanisation-population-boom-lakes-urban-jungle-dead-city-2765760.html.

Drakakis-Smith, David, Tanya Bowyer-Bower, and Dan Tevera. 1995. "Urban Poverty and Urban Agriculture: An Overview of the Linkages in Harare." *Habitat International* 19 (2): 183–93. https://doi.org/:10.1016/0197-3975(94)00065-A.

Fernandes, Leela, and Patrick Heller. 2006. "Hegemonic Aspirations: New Middle Class Politics and India's Democracy in Comparative Perspective." *Critical Asian Studies* 38 (4): 495–522. https://doi.org/:10.1080/14672710601073028.

Frazier, Camille. 2018. "'Grow What You Eat, Eat What You Grow': Urban Agriculture as Middle Class Intervention in India." *Journal of Political Ecology* 25 (1). https://doi.org/10.2458/v25i1.22970.

Galvin, Saila Seshai. 2018. "The Farming of Trust: Organic Certification and the Limits of Transparency in Uttarakhand, India." *American Ethnologist* 45 (4): 495–507. https://doi.org/10.1111/amet.12704.

Government of India. 2011. 2011 Census. Accessed June 1, 2018. http://www.censusindia.gov.in/DigitalLibrary/Books.aspx.

Guthman, Julie. 2004. *Agrarian Dreams: The Paradox of Organic Farming in California*. Berkeley: University of California Press.
Guthman, Julie. 2008. "Bringing Good Food to Others: Investigating the Subjects of Alternative Food Practice." *Cultural Geographies* 15 (4): 431–47. https://doi.org/10.1177/1474474008094315.
Hite, Emily Benton, Dorie Perez, Dalia D'ingeo, Qasimah Boston, and Miaisha Mitchell. 2017. "Intersecting Race, Space, and Place through Community Gardens." *Annals of Anthropological Practice* 41 (2): 55–66. https://doi.org/10.1111/napa.12113.
K. Bhumika. 2016. "Bengaluru's Growing Pride." *The Hindu*, February 19, 2016. Accessed October 1, 2020. https://www.thehindu.com/features/metroplus/events/bengalurus-growing-pride/article8296120.ece.
Menezes, Naveen. 2016. "Bengaluru Will be an Unliveable, Dead City in 5 Years." *Deccan Herald*, May 2, 2016. Accessed October 1, 2020. https://www.deccanherald.com/content/543880/bengaluru-unliveable-dead-city-5.html.
Mintz, Sidney W., and Daniela Schlettwein-Gsell. 2001. "Food Patterns in Agrarian Societies: The "Core-Fringe-Legume Hypothesis" A Dialogue." *Gastronomica* 1 (3): 40–52. https://doi.org/10.1525/gfc.2001.1.3.40.
Nagendra, Harini. 2016. *Nature in the City: Bengaluru in the Past, Present, and Future*. New Delhi: Oxford University Press.
Nair, Janaki. 2000. "Language and Right to the City." *Economic and Political Weekly* 35 (47): 4141–46.
Nair, Janaki. 2005. *The Promise of the Metropolis: Bengaluru's Twentieth Century*. New Delhi: Oxford University Press.
Poulsen, Melissa N., Kristyna R. S. Hulland, Carolyn A. Gulas, Hieu Pham, Sarah L. Dalglish, Rebecca K. Wilkinson, and Peter J. Winch. 2014. "Growing an Urban Oasis: A Qualitative Study of the Perceived Benefits of Community Gardening in Baltimore, Maryland." *Culture, Agriculture, Food and Environment* 36 (2): 69–82. https://doi.org/10.1111/cuag.12035.
Pudup, Mary Beth. 2008. "It Takes a Garden: Cultivating Citizen-Subjects in Organized Garden Projects." *Geoforum* 39 (3): 1228–40.
Rademacher, Anne. 2015. "Urban Political Ecology." *Annual Review of Anthropology* 44 (1): 137–52. https://doi.org/10.1146/annurev-anthro-102214-014208.
Rademacher, Anne M., and K. Sivaramakrishnan, eds. 2013. *Ecologies of Urbanism in India: Metropolitan Civility and Sustainability*. Hong Kong: Hong Kong University Press.
Rademacher, Anne, and K. Sivaramakrishnan, eds. 2017. *Places of Nature in Ecologies of Urbanism*. Hong Kong: Hong Kong University Press.
Ranganathan, Malini. 2011. "The Embeddedness of Cost Recovery: Water Reforms and Associationism at Bangalore's Fringes." In *Urban Navigations: Politics, Space, and the City in South Asia*, edited by Jonathan Shapiro Anjaria and Colin McFarlane, 165–90. New York: Routledge.
Simatele, Danny Malula, and Tony Binns. 2008. "Motivation and Marginalization in African Urban Agriculture: The Case of Lusaka, Zambia." *Urban Forum* 19 (1): 1–21. https://doi.org/:10.1007/s12132-008-9021-1.
Slocum, Rachel, and Kirsten Valentine Cadieux. 2015. "Notes on the Practice of Food Justice in the US: Understanding and Confronting Trauma and Inequity." *Journal of Political Ecology* 22: 27–52.

Sokolovsky, Jay. 2011. "Civic Ecology and the Anthropology of Place: Urban Community Gardens and the Creation of Inclusionary Landscapes. *Anthropology News* 52 (3): 6. https://doi.org/10.1111/j.1556-3502.2011.52306.x.

Solomon, Harris. 2015. "Unreliable Eating: Patterns of Food Adulteration in Urban India." *BioSocieties* 10 (2): 177–93. https://doi.org/10.1057/biosoc.2015.10.

Srinivas, Smriti. 2001. *Landscapes of Urban Memory: The Sacred and the Civic in India's High Tech City*. Minneapolis: University of Minnesota Press.

Srinivasan, Sarayu. 2016. "No, Bengaluru Won't Be 'Dead' in 5 Years, but Things Are Pretty Messed Up for the City." *The News Minute*, May 4, 2016. Accessed October 1, 2020. https://www.thenewsminute.com/article/no-bengaluru-wont-be-dead-5-years-things-are-pretty-messed-city-42642.

Thakur, Aksheev. 2017. "In Three Years, Bengaluru Will Be a 'Dead City,' Says IISc Study." *Deccan Chronicle*, August 27, 2017. Accessed October 1, 2020. https://www.deccanchronicle.com/nation/current-affairs/270817/in-three-years-bengaluru-will-be-a-dead-city-says-iisc-study.html.

The Times of India. 2015. "Poison Your Platter: Even the Veggies You Eat May Be Unsafe." June 9, 2015.

Tracy, Megan. 2010. "The Mutability of Melamine: A Transductive Account of a Scandal." *Anthropology Today* 26 (6): 4–8. https://doi.org/10.1111/j.1467-8322.2010.00768.x.

Unnikrishnan, Hita, and Harini Nagendra. 2014. "Privatizing the Commons: Impact on Ecosystem Services in Bangalore's Lakes." *Urban Ecosystems* 18 (2): 613–32. https://doi.org/10.1007/s11252-014-0401-0.

Yan, Yunxiang. 2012. "Food Safety and Social Risk in Contemporary China." *The Journal of Asian Studies* 71 (03): 705–29. https://doi.org/10.1017/S0021911812000678.

3

The Village at the End of the World: Ecologies of Urbanism in Climate Crisis Imaginaries

Kasia Paprocki

> You see this? This is nothing. It's not anything you can make an enterprise out of. This is nothing.

We were standing on the edge of a crumbling embankment in the Sundarbans of West Bengal, staring out at a potato field. It was late January, the tail end of the potato harvest, and the land was dotted with small piles of the remaining yield.[1] The speaker was gesturing animatedly at the field, with his gaze turned away from it and toward us. A few farmers moved about in the field with little attention to us, or indeed this somewhat brazen dismissal of their work (pronounced in English for the benefit of the gathered audience peering down from the embankment).[2]

This was day two of a sightseeing junket of the Indian side of the Sundarbans. The world's largest mangrove forest, the Sundarbans straddles the border of India and Bangladesh, flanked to the south by the Bay of Bengal. The trip had been planned and organized by the World Wide Fund for Nature-India (WWF) through a program supported by the World Bank. In addition to WWF staff, my companions were journalists, donors, and government officials from Bangladesh and India. The WWF program officials were on a mission to reveal to this collective not only the unique ecological characteristics of the region but also a vision of the region's future that they sought to promote through a new climate change adaptation program they were in the process of expanding. It entailed planned retreat from coastal villages and associated urban development to accommodate the climate migrants such a transformation would produce. In this vision, climate change adaptation would

---

1. Excerpts from this chapter have appeared previously in Paprocki 2020, "The Climate Change of Your Desires: Climate Migration and Imaginaries of Urban and Rural Climate Futures." *Environment and Planning D: Society and Space* 38 (2): 248–66. I thank *Environment and Planning D: Society and Space* for permitting the reuse of the material in this chapter.
2. This chapter is based on research findings derived from over two years of multisited ethnographic research conducted primarily in 2014–2015. This research involved interviews and participant observation in rural communities in coastal Bangladesh and with migrants from those communities in peri-urban Kolkata, as well as participant observation with development practitioners and policy makers in Dhaka, Bangladesh, and Kolkata, India.

require the active destruction of rural futures in order to forge new, resilient, and prosperous urban ones.

In what follows, I outline two different normative imaginaries of life in the Sundarban region in the time of climate change. One imaginary embraces the primacy of forms of life including shrimp and middle-class tourism from nearby cities, while the other values agrarian livelihoods of existing rural populations. The tension between these two imaginaries is brought to the fore in narratives of climate change adaptation through urbanization and export-led growth. Narratives and interventions described in this chapter indicate a shifting landscape of the forms of life considered suitable for a climate-changed future in this region. I examine the emergence of new ecologies of both rural and city space, and ongoing contestations over these imaginaries.

I explore this emergence of new city and rural ecologies through a series of projects of dispossession in coastal communities in Bangladesh and India, surrounding the Sundarbans. This involves narratives of climate crisis and adaptation and accompanying development interventions, paired with an examination of the political economy of rural out-migration from this region to Kolkata. I illuminate the intersections of three sites: a village in coastal Bangladesh, a slum on the outskirts of Kolkata, and the Indian Sundarban region being targeted for planned retreat in anticipation of climate change. I examine how climate change becomes the ecological and temporal context within which new models of development are imagined for the present and future not only in this region, but throughout the rest of the world (cf. Zeiderman 2016). In this sense, an investigation of these particular urban climate imaginaries offers a window into the biopolitical governance of life under climate change more broadly. Here, the political economy of development, climate change, and the rural-urban transformations intersect to shape and be shaped by spatially interconnected modes of governing in anticipation of an uncertain future.[3] This emerging biopolitics of climate change suggest both new interventions and new rationalities of intervention governing the work of making live and letting die in the time of climate change (Foucault 2003; Li 2010).

In service of this analysis, I engage recent literatures that examine new imaginaries and materialities of life in a climate-changed future. Specifically, I draw on insights from Anna Tsing's and Donna Haraway's studies of living in the ruins of capitalism and climate change in order to reflect on ongoing transformations in interconnected human and nonhuman ecologies (Haraway 2016; Tsing 2015). Here I refer to plural *ecologies*, as multiple and shifting ways of knowing nature. New ecologies of the rural forged by development agencies in this region imagine non-agrarian futures for these rural areas that have long been home to farmers and farming communities. Shrimp, tigers, and tourists from nearby cities are the

---

3. See Goh (2020) for a similar approach to examining climate change adaptation through global-urban networks.

forms of life that populate these imagined future ecologies. For these forms of life and imagined ecologies to thrive requires the dispossession of the rice and potato farmers who have long inhabited these rural spaces. This imaginary gives rise to new city ecologies, as well, and these new ecologies promise alternative livelihood opportunities to the dispossessed rice and potato farmers.

I understand these processes as collectively constituted within what I have elsewhere called an *adaptation regime*, a socially and historically specific configuration of power that governs the landscape of possible intervention in the face of climate change (Paprocki 2018). In this chapter I investigate the epistemic and material dynamics through which the adaptation regime promotes a vision of transition away from agrarian livelihoods toward export-oriented production based in cities, necessitating rural decay for the sake of urban expansion. Does a vision of urban climate resilience require the devaluation of rural lives and livelihoods? The broader implications of these investigations are to argue that we cannot understand the dynamics governing the production of urban natures without close attention to the associated production of rural natures.

Moral Ecologies of the Future

We might consider these new logics and practices an emerging moral ecology of climate change in the Sundarbans, drawing on Anne Rademacher's articulation of moral ecology as collective understandings of what is "right and necessary for the good of society and the environment" (Rademacher 2018, 16). Yet, like moral economies (Wolford 2005), moral ecologies are multiple and shifting. Collective understandings of necessary ecological formations and transformations both shape and are shaped by systems of power at multiple scales and are actively forged and contested in the present.[4] This moral ecology produced and governed under the adaptation regime is embedded in complex systems of power through which new forms of life are forged for Asian cities and villages in the time of climate change. Where do urban moral ecologies intersect with rural ones, and where do they elide them?

The demise of imaginaries of rural futures through these new moral ecologies is significant not only because of the changes it facilitates, but also because of the alternative futures these moral ecologies omit. For several decades, social movements led by farmers in this coastal region have mobilized to defend continued agricultural production in their villages, resisting a transition toward commercial shrimp aquaculture and the agrarian dispossessions it entails (Adnan 2013; Paprocki 2019). Today, these movements continue to gain traction, not only supporting continued rice production but also championing a return to rice agriculture in communities

---

4. See also Elliott's examination of competing moral economies of flood risk and climate change adaptation (2017).

that had earlier transitioned to shrimp (Afroz, Cramb, and Grünbühel 2017; Paprocki 2019). Their alternative visions of the persistence of agrarian futures contrast sharply with visions and discourses of agrarian decline. The growing success of these movements speaks to the possibility of continued life and agricultural production in this region under climate change, undermining notions of the inevitability of ecological crisis and the erasure of agrarian livelihoods.

At odds with these local visions, there is historic precedent for thinking about the Sundarbans as a zone of social and ecological backwardness, demanding exceptional modes of governance. Bhattacharyya (2018a, b) has documented how the British East India Company and the Raj that followed used a variety of legal, bureaucratic, and engineering technologies to attempt to tame a landscape that was fundamentally resistant to administrative control. A sense of the region's ecological vulnerability has been shaped historically by aggressive dynamics of artificial land reclamation and resettlement dating to the colonial period, combined with a proximity to Bangladesh that has led to more recent habitation by migrants considered undesirable from this marginal neighbor (Iqbal 2010; Samaddar 1999). As Harms describes in this volume, such unruly ecologies are often linked with ascribed moral qualities of their inhabitants. In Bengal, this colonial moral ecology has continued to shape normative ideas about the agrarian lives and livelihoods of people living near the Sundarbans.

Along with these social and physical dynamics, the Sundarbans are a unique mangrove ecosystem, home to several rare and endangered species, including the Bengal tiger. As Jalais has written, "throughout the recent history of the Sundarbans, the very presence of people in the region has been seen as a hindrance to its development as a 'natural' haven for wildlife" (Jalais 2010, 9). Moreover, the development and expansion of cities in this region has historically entailed the active devaluation of rural space, enrolling it in dynamics of capital accumulation (Bhattacharyya 2018b). Yet, climate change has created an opportunity for new modes of moral and material governance of the region. Instead of benign neglect and underdevelopment, the notion that adaptation in this area should be carried out through actively dismantling its social and physical infrastructures has suggested new opportunities for regional and national growth and accumulation, now repackaged as innovations in climate change adaptation. In the final section of this chapter, I examine how this is now taking place through an explicit vision of planned retreat and active devaluation of rural futures.

Examining these moral ecologies for cities and village spaces in Bengal offer an important opportunity for understanding the politics of futurity in the Asian city today. I refer here to Asia as both site and symbol, drawing on Ananya Roy's provocative question, "When is Asia?," through which she considers the project of imagining Asia not as a geographically bounded location but as a historical conjuncture and "a set of citationary relations through which a politics of futurity is crafted" (Roy 2016, 317). New moral ecologies of the climate-resilient Asian City are celebrated by development practitioners, donors, and policy makers, as exemplified by programs supported by the Rockefeller Foundation (Tanner et al. 2009), USAID (Lowry, Fuchs, and Bettinger 2015), UN-Habitat (UN-Habitat and UN ESCAP 2014), and the Asia Foundation (Patel 2015), which collectively conceptualize the Asian City as the frontline of experimentation with new modes of governing in the time of climate change (Reed et al. 2015; Tyler and Moench 2012). As elsewhere in the world, discourses on new paradigms of sustainable development and low-carbon living tend to foreground high-density lifestyles in cities (Wachsmuth, Cohen, and Angelo 2016). Climate change becomes the ecological and temporal context within which new models of development are forged for the present and future, not only in the Asian City but throughout the rest of the world (cf. Zeiderman 2016).

Even as these disparate sites across rural and urban Bengal illuminate processes taking place in the context of climate change across this region, they also highlight the different political economies of development in which such moral ecologies intervene on either side of the border. In India and Bangladesh, people are situated in structures through which they can aspire in different ways. The political economy of rural-urban migration in Bangladesh is different from that of India, where opportunities for migration and urban livelihoods are expanding at a different pace (even as these opportunities are inequitably distributed within those countries). These differences illuminate the contradictions of imagining biophysical and social transformation proceeding evenly on either side of the border. They also highlight the problems with pursuing such transformations without attention to the unique political economies of development in any community in which adaptation takes place.

Theorizing the End of the World

The competing moral ecologies of the future examined in this chapter point to different social understandings of ecological gain, loss, and threats in the time of climate change. Each respond to threat and disturbance through different normative visions of possibility for the future. Collectively, they complicate progress narratives by exposing tensions between ideas of progress for different people and different forms of life that the visions unevenly celebrate. As Tsing tells us, "deciding what counts as disturbance is always a matter of point of view" (Tsing 2015, 161). Indeed, progress in one vision is a disturbance for another, and vice versa. In these communities in Bangladesh and West Bengal, we find that the ecologies of urbanism promoted by development practitioners entail disturbance of agrarian lives and livelihoods in order to bring capitalist development more fully to life in the city.

We need to be attentive to the power relations embedded in this celebration of alternative forms of life (and death) in the time of climate change. In *Staying with the Trouble*, Haraway encourages openness to these alternative imaginaries, which she describes as learning to live with spiders, monsters, and "snaky, unheroic, tentacular, dreadful ones" (Haraway 2016, 43). Yet, like mushrooms, these dreadful ones do not necessarily exercise political agency. "They do not engage in resistance or refusal," critics have noted (Padwe 2019, 436). So, while we can recognize the celebration of nonhuman life in the ecologies of urbanism, we need to not let this celebration obscure the dynamics of dispossession.

Attention to peasant moral ecologies challenging a globally hegemonic neoliberalism has often been disparaged as agrarian romanticism. Indeed, in the modernist imaginaries of climate change adaptation producing new configurations of city space, the loss of agrarian livelihoods is a necessary and also normatively progressive corollary. In this sense, climate change adaptation reproduces the telos of capitalist modernity. In working to examine the places of nature within and beyond the city, we find not only multispecies entanglements but also the conditions of possibility for other ways of living. By this I don't mean utopia ("this is no place to search for utopia," Tsing tells us [2012, 152]); rather, as Collard and colleagues describe, "futures with more diverse and autonomous forms of life and ways of living together" (Collard, Dempsey, and Sundberg 2015, 323). These analytical rubrics direct our attention not only to the ways in which certain kinds of urban imaginaries deem forms of life beyond the city disposable, but they also open up space for us to imagine otherwise.

Kolanihat: Moral Ecologies of Agrarian Dispossession

Kolanihat[5] is a village in Khulna's Paikgachha subdistrict, about twenty miles south of Khulna City as the crow flies, and five miles from Paikgachha town, the nearest trading market. Investigating recent transformations in production and social reproduction in Kolanihat offers a window into the overlap of Khulna's rural communities with the region's larger political economy of development. Until the mid-1980s, most residents of Kolanihat produced one or two agricultural crops per year, the most important of which was *aman* (monsoon season) rice. This limited growing season was the result of the low land elevation that kept some of the village's fields under water for much of the year. Nonetheless, the fertile alluvial soils enriched by sediment deposits from the floodwaters of the adjacent river facilitated an abundant crop, and most residents report historic surplus production that kept their families fed throughout the year. While many of the village's residents did not own land, most were engaged in agricultural production (a survey from a nearby village in 1987 found that over 50 percent of residents were either landless or marginal land holders, requiring them to sharecrop or sell their labor for seasonal agricultural production [A. K. Datta 1998, 31]). At this time, the landless and land poor

---

5. I have changed the names of villages and people in this section in order to protect the identities of my informants.

survived on various combinations of sharecropping, day laboring, and seasonal labor out-migration.[6]

In the 1980s, Bangladesh was undergoing a period of rapid structural adjustment. Along with the growth of the country's garment industry, the expansion of commercial production of saltwater shrimp was identified as a key strategy in efforts toward export diversification. Kolanihat was enrolled in this expansion in 1986, when Wakil, a wealthy businessman from Khulna City chose it as a spot for investing in a large *gher*[7] (shrimp cultivation) operation. Throughout the region at this time, huge tracts of land were being converted into *ghers* from rice farming lands through varying degrees of consent from local communities. While some lands were leased from their owners (who often found later on that their use went unpaid or underpaid), many were forcibly taken through illegal and often quite violent land grabbing. In 1990, in a nearby village just across the river from Kolanihat, Karunamoyee Sardar, a local farmer and landless movement leader, was abducted and murdered in the midst of a protest against land grabbing for shrimp cultivation in her village. Narratives about the relative use of force to compel this transition in Kolanihat differ. While many of the village's landless residents tell stories about armed guards hired by Wakil either to force the land grab or to prevent theft from the *ghers* once they were established, wealthier residents tell stories about a calmer process through which they agreed to lease out their lands, only to find later that they were not paid as agreed, were paid less than expected, or were unable to easily reclaim the use of their lands at the end of the lease term.

Meanwhile, Wakil built his own sluice gate to bring salt water from the river into his *gher*, effectively allowing him to control the management of water within most of Kolanihat and the surrounding area. This control over the water management within the village has serious implications for life within Kolanihat. While Wakil's sluice allows for the uninhibited flow of water between the river and the land he controls, residents describe it as just one element within a larger political system in which the financial capital and political influence from nearby cities comes to shape both their physical landscape and their ability to survive within it. One man told this story about the ongoing struggles to keep shrimp cultivation out of the village,

> There have been clashes with them [the businessmen who own *ghers*]. These people live in the city, some live in Khulna, Satkhira.[8] The rich people who control the [local] administration have been torturing us. We repair the river embankments and then they come at night with the police and they break them down again [to allow the inflow of salt water]. When we go out in the morning they send goons hired from the city to attack us. They torture us. If we try to go to the police station

---

6. This labor out-migration was largely limited to male landless farmers, where their wives and families stayed in the village year-round (see also Paprocki and Cons 2014).
7. *Gher* is the word used for the large saltwater ponds or bogs used for aquaculture cultivation.
8. Khulna and Satkhira are the two largest cities in Khulna Division, where much of the shrimp trade is based.

[to file complaints], they make us file a General Diary and they say "we will look into it." They say they will look into it but that very night the water is released into the *gher* again.

This man's testimony offers a window into the rural political economy of shrimp production and its urban interconnections. He describes how agricultural production in the village has been subverted by the economic interests of outsiders and how the complicity of local authorities has actively sustained this subversion. He describes how these power dynamics are physically inscribed into the landscape of the village, most clearly through struggles over the protective embankments that keep the salt water out (or in).

This power also reshapes the internal landscape of the village, where the fertility of the soil, increasingly salinated, deteriorates. When the salt water is brought in from the river, it fills the *ghers* and seeps into the surrounding farmland, such that it becomes impossible to farm rice in adjacent plots. Gradually, the salinity has killed the trees in the village, crept into homesteads, and made it virtually impossible to cultivate the small garden plots that support subsistence consumption throughout rural Bangladesh. As agriculture has given way to aquaculture, the local labor market has also transformed dramatically. Residents of Kolanihat estimate that shrimp aquaculture requires somewhere between 1 percent and 10 percent of the amount of labor as rice agriculture requires.[9] Thus, this shift has resulted in a significant labor surplus in the village, a change experienced most seriously by the significant proportion of landless laborers in the village who depended on this work for their survival. These people have been forced to migrate out of Kolanihat to find work, many permanently.

Many of these recent changes in Kolanihat can be understood in relation to the transformations in emerging social imaginaries of life in the time of climate change (Paprocki 2019). The water logging and soil salination caused by the inflow of salt water for shrimp cultivation have been frequently attributed to the results of climate change and sea level rise by journalists, development practitioners, and even some academics (Brammer 2014; Szczepanski, Sedlar, and Shalant 2018). Consequently, shrimp aquaculture has been proposed as a climate change adaptation strategy by many within the development and donor communities in Bangladesh who suggest that the use of salinated and waterlogged former agricultural lands for shrimp is a logical and lucrative adaptive response to the current ecological crisis (Paprocki 2018).

Finally, the migrations resulting from this process of depeasantization have been reframed as climate migration (Norwegian Refugee Council 2015; Shamsuddoha et al. 2012), obscuring the real dynamics of agrarian change in the region and their

---

9. There is no clear consensus in the academic literature on this discrepancy in labor requirements between rice and shrimp. Belton's research (2016) indicates a less dramatic, but nevertheless serious shift in labor demand for shrimp, citing a requirement of 54 percent more labor for rice agricultural systems relative to shrimp production.

consequences (Brammer 2009). The cascading impacts of these "climate migrations" have been hailed as among the greatest global security threats of the twenty-first century, with out-migration to India from these low-lying islands in coastal Khulna cited as a particular flashpoint of climate vulnerability. John Podesta (formerly chief of staff to Bill Clinton) and Peter Ogden of the Center for American Progress write that "India will struggle to cope with a surge of displaced people from Bangladesh, in addition to those who will arrive from the small islands in the Bay of Bengal that are being slowly swallowed by the rising sea" (Podesta and Ogden 2007, 117), explaining that "these desperate individuals go where they can, not necessarily where they should" (131). Certainly, the question of where migrants from Khulna *should* go is shaped by emerging moral ecologies of climate change, both in this region as well as globally.

As the physical landscape and labor market in Kolanihat transform, so too do migration patterns of its residents. Dwindling agricultural labor opportunities force those who previously relied on sharecropping and seasonal day labor in agriculture to leave in search of more durable sources of income. At the beginning, this process involved moments of violence and it has been punctuated by incidents of violent dispossession throughout. Yet, over the past several decades, the secular dynamics of depeasantization have turned slow and less conspicuous. Many who previously relied on seasonal migration have been forced to leave more permanently (Paprocki 2019). Some find jobs in brick manufacturing in peri-urban areas around Bangladesh. Some go to Khulna City, where there are jobs in construction as well as in de-heading shrimp in factories where it is then frozen for export. Some find work in construction or garment manufacturing in Dhaka. Yet, residents say that most who leave Kolanihat travel across the border to Kolkata. Some do so on a seasonal basis, but many leave permanently and bring their families with them.

One resident of Kolanihat described to me this slow process of dispossession; he formerly worked as a day laborer, the income from which supported his family, supplemented by a robust garden plot in their homestead. Several years ago, he was injured in an accident and took a microcredit loan of 5,000 taka (about US$60) to pay for the associated medical expenses. With insufficient earning opportunities in the village while he recovered, he struggled to repay the loan and his debt grew. He traveled once to Kolkata and found that work was available there that would support his livelihood more sustainably. Within several years, his debt had grown to 17,000 taka (about US$203). At that point, the debt had become insurmountable, and he could see no viable future livelihood in Kolanihat.[10] In September 2014, he told me he planned to sell everything and leave for Kolkata permanently. Migrations like this one are a prominent feature of the political economy of the development of shrimp aquaculture in Kolanihat. While the mechanism of dispossession is less conspicuous than the violence of an overt land grab, its impacts on the population

---

10. This reflects a pattern of cyclical debt and dispossession through microcredit observed elsewhere in rural Bangladesh (Paprocki 2016).

of the village have been immense. Describing these vast migrations from her village to Kolkata, another woman in Kolanihat explained to me, *"jibika nirbhor kore jay"* ("they leave as their livelihoods depend on it").

New Town: Moral Ecologies of Urban Migration

When Kolanihat's migrants travel to Kolkata, most go to a small enclave on the outskirts of the burgeoning satellite city of New Town.[11] New Town has been planned for residential use and as a hub of Kolkata's growing IT sector—now envisioned as a new mode of greening urban development. It has also been the site of a battle between competing visions of urban green growth in India (Das Gupta 2017). While Prime Minister Narendra Modi envisioned New Town as a key site in his "Smart City" mission for sustainable urban development, Chief Minister of West Bengal Mamata Banerjee has sought to develop New Town as India's first "Green City."[12] While these visions reflect substantive differences between India's major BJP and Congress parties over equity in urban development, water rights, and centralization of the planning process (Ghoshal 2016), both require the labor of migrants in service of their expansion.

Migrants from what is now Bangladesh have been traveling to this part of greater Kolkata since Partition, when the area was still largely farmland. Thus, today many recent Bangladeshi migrants rent space from wealthier, more established migrants who have been there for decades. Roy has referred to such spaces surrounding greater Kolkata as the "rural-urban interface," by which she suggests not only the spatial proximity of the rural and the urban but also their interconnected political economies (Roy 2003). I explore here how their liminal status between urban and rural can be understood through the relationship of these migrants to the rural spaces from which they have come as well as through their relationship to the city they inhabit and are helping to construct. The migrants refer to this space where they live as *gram*, meaning "village," denoting the apparent rural geographical imaginaries through which they construct this space (cf. Jazeel 2018). Yet, the spatial configuration of the community looks more like an urban slum (or *bosti*) than it does like the rural villages from which they have come. The small dwellings made of corrugated metal and cinder blocks are tightly squeezed together, with some perched precariously on bamboo stilts over an open sewage canal. Instead of socializing in the spacious open courtyards of the traditional Bangladeshi village, social interactions are squeezed between narrow pathways or spill out into the surrounding area of New Town, into parks, bus stands, and sitting in the grass around the large holding basin of a water treatment facility.

---

11. I avoid naming the specific neighborhoods inhabited by Bangladeshi migrants in New Town to protect the identities of my informants.
12. For more on urban future imaginaries in India's "Smart Cities" initiative, see Datta (2019).

While these migrants blend into the urban space in some ways quite inconspicuously, New Town has also been planned in many ways to actively exclude them. Large walls separate the *gram* from the impressively large developments that house the community's wealthier residents. The names of these buildings displayed prominently on many of their facades reveal the future imaginaries of their inhabitants. As we walked together past buildings called "Website Housing" and "TechnoNest," one young migrant from Kolanihat explained to me that it is difficult to find domestic work in these homes because their inhabitants want to see documentation of legal status in India from prospective domestic staff. However, the labor market where day laborers are recruited for construction of these buildings hosts Bangladeshi migrants almost exclusively. Compared to garment work in Dhaka, these construction jobs pay much more for almost half of the working hours (depending on one's level of skill), so he finds that Kolkata offers the opportunity for a more comfortable lifestyle than Dhaka.

Other migrants from Kolanihat expressed a similar kind of ambivalent belonging in Kolkata. Some say they don't like it there and don't want to stay and would prefer to go home. This comparison, in which they convey longing for the declining agricultural livelihoods of their rural homelands, was the emotion these migrants articulated to me most commonly. One woman explained, "I like it here OK, but it's not like Bangladesh. There's not enough work in Bangladesh, but it's better there than anywhere else." I heard these sentiments repeated again and again from migrants in New Town. While some had come very recently, others had been there for twenty years or more. The earlier migrants were landless people who previously relied on day labor or sharecropping but found that the shrimp boom created an insufficient number of jobs to keep them employed. One such migrant told me that every landless person in her village had ultimately migrated here to New Town. More recent migrants were smallholders, some of whom had participated in shrimp production but hadn't found it to be profitable enough to survive on or who experienced some kind of personal or familial crisis that forced them to leave. Many were the sons of smallholders who continued to cultivate shrimp but who were struggling or didn't see a viable future for it.

In general, these were people who continued to identify deeply with the villages from which they had come and the peasant livelihoods they led there. Even as they had moved to New Town, the home of Kolkata's future imaginaries in both a material and ideological sense, they continued to very actively value, embrace, and identify with rural lives and livelihoods. This identification is precarious in a context in which not only their present livelihoods depended on this urban political economy but also the possibility of a rural future, farming rice in the villages from which they had come, was not guaranteed. In the narratives of these migrants, their aspirations for rural futures coexisted with their active participation in the construction (both literal and imagined) of urban futures.

Sundarbans: Moral Ecologies of Imagined Erasure

Here we return to the WWF sightseeing junket where this chapter began. "This is nothing," says the WWF official, motioning toward the potato field. Indeed, the discursive erasure of this rural space is a key component of the production of urban climate imaginaries in Kolkata. The devaluation of agrarian livelihoods, in the formulation of this official, is seen as necessary to imagining a more desirable urban future. It is constitutive of a new moral ecology of climate change that shapes both rural and urban livelihoods across this region.

This vision of climate futures is spelled out more directly in a policy brief published by WWF in 2016, titled "Away from the Devil and the Deep Blue Sea: Planned Retreat and Ecosystem Regeneration as Adaptation to Climate Change" (Ghosh et al. 2016). In this report, a group of academics and development practitioners working for WWF articulate a plan for the implementation of planned retreat from the Sundarbans, and they explain the economic benefits that would derive from such a transformation. This vision of retreat builds on a growing discussion in both policy and academic communities concerning the possibility of planned relocation of communities as a climate change adaptation strategy (Koslov 2016, 2019; Marino 2018). The report extends these discussions to a concrete, empirical investigation comparing the value of the existing agrarian political economy to an alternative vision of planned retreat in which agriculturalists in these coastal villages relocate to "newly developed areas in [a] nearby stable zone" (Ghosh et al. 2016, 12), where they will find work in the service sector and "skilled employment" (meaning outside the agricultural sector). The normative values underpinning this bold vision are part of a distinctive moral ecology of climate change not only for the Sundarbans but for Kolkata, all of India, and beyond (Farbotko 2010; Pulido 2018).

The report describes this vision of social and ecological transformation for planned retreat proceeding in four phases, culminating in the year 2050. The plans are both material (relating to technical and economic interventions) as well as explicitly epistemic (relating to the kinds of work that will need to be done to reshape desires and imaginaries of life in the time of climate change). In Phase I, the "high vulnerability zone" would be demarcated, and a policy framework implemented to prevent "outsiders" from moving into the area. There are two significant implications of this: the first is the creation of barriers to migration by Bangladeshis, who are thought to be disproportionately represented among inhabitants of this Sundarban region. The second is that by creating impediments to migration and land acquisition, the land in the region would be effectively taken out of circulation, and thus economically devalued.[13]

In Phase II, new physical infrastructure is built in the "stable zone" meaning development in Kolkata and other urban or peri-urban areas. Some physical

---

13. For more on the dynamics of devaluation in the context of climate change, see Knuth (2017) and Elliott (2019).

infrastructure costs associated with this phase cited in the report's appendix include the establishment of Industrial and Information Technology Training Institutes. With this in mind, it becomes clear that the report's references to the costs of "reskilling" are a metonym for the costs of transforming rural futures into urban ones. These material interventions in Phase II are accompanied by explicit epistemic interventions in Phase III, which involves "preparing the residents for this change in order to minimise their psychological barrier towards the movement from the vulnerable to the less vulnerable zone" (Ghosh et al. 2016, 12). The report specifies that at this stage resettlement is undertaken by choice, noting "the movement is envisaged as voluntary and 'organic'" (Ghosh et al. 2016, 12). Yet, even in the absence of forced relocation, the "choice" to migrate in this context is undertaken within extremely constrained conditions of the active erasure of livelihood possibilities and devaluation of the assets that make these agrarian livelihoods possible. These manufactured constraints on migration choices are thus a more explicit (yet perhaps logical extreme) of the rural-urban migration choice facing residents of Kolanihat today, for whom the political economy of shrimp production offers no viable rural future. This political economic transformation might thus be seen as the "adaptive" precursor readying the ground for the emergent strategy of planned retreat.

In the final phase, remaining residents are relocated (presumably by force, although the report does not use this language, insisting on the importance of framing the process as benign). Once the lands in this "high vulnerability zone" have been entirely depopulated, they will be made available for mangrove regeneration. As described in the report, the benefits of this transition away from an agrarian political economy are manifold. In addition to the benefits of storm surge protection and carbon sequestration facilitated by mangrove reforestation, they describe a range of economic opportunities opened up. These include crab and fishery production, the collection of honey and prawn larvae (for use in aquaculture), and new revenues from tourism amongst the growing population of nearby cities. These tourist possibilities were highlighted in particular on the sightseeing junket through visits to existing eco-resorts catering primarily to middle-class visitors from Kolkata. The analysis of the report's authors suggests that collectively the benefits of these alternative income streams would be 12.8 times greater than the economic benefits derived from the current agrarian political economy in the region. The report thus offers a systematic vision not only of the process of planned retreat but of a plan for combined material and epistemic interventions to facilitate a transition from rural to urban climate futures. The demise of these agrarian futures is framed as necessary to the achievement of this alternative vision of (urban) development in the time of climate change.[14]

---

14. For a corollary discussion of alternative urban climate imaginaries, see Cohen (2016) and Goh (2017).

Conclusion

How do we imagine a desirable climate future? What spatially differentiated processes of enclosure and emergence are entailed in that imaginary? In this chapter I have traced the links between three sites, both rural and urban, and the interconnections between the competing moral ecologies that shape them. In the process I have mapped the relationship between the managed decline of rural futures and the development of new futures in cities. The devaluation of the lives and livelihoods that currently inhabit this rural space is fundamental to the planning process through which this urban future is operationalized. The vision for a modern Kolkata requires the labor of rural migrants and the dystopic imagination of the impossibility of a future for the communities from which they have come.

Attention to posthuman ecological imaginaries such as those of Haraway (2016) and Tsing (2015) directs us to new ways of valuing the nonhuman lives and ecologies that might emerge in the ruins of these agrarian livelihoods. Yet, they also suggest new spaces for deeply human political projects. While Haraway urges attention specifically to new frontiers in how to think about the coexistence of human and nonhuman life on earth, her analysis also directs us to alternative moral ecologies that suggest different ways to organize existing life on earth, the human aspirations around which we govern ourselves, and the teleologies that shape our ideas about progress and the future. In Bengal, this might mean a renewed openness to agrarian livelihoods not only as a means of persistence but within a new ecology of urbanism that doesn't require the death of lives and livelihoods beyond the city.

The ecologies of urbanism project asks us to interrogate how certain configurations of life and livelihood, both human and nonhuman, are made to thrive in transforming geographies of city space, while other configurations are deemed disposable. These imaginaries are shaped by existing relations of power in both rural and urban space. As Rademacher and Sivaramakrishnan write,

> Our analytic, that of urban ecologies, assumes the presence of multiple, simultaneous, and overlapping representations of the urban nature-urban culture interface. Each represents competing visions, ideas, and stakes of urban environmental change. Their corresponding efforts to ensure, create, or imagine ecological stability are often infused with, and shaped by, aspirations for political, social, or cultural stability; to promote particular urban ecologies may also involve the reproduction or contestation of cultural ideas of belonging to certain social groups, including the city, the nation-state, the region, and the realm called the "global." (2013, 11)

The competing moral ecologies of both rural and urban spaces in Kolanihat, New Town, and the Sundarbans demonstrate how climate futures are shaped by political economic aspirations that privilege some forms of life above others. Discourses of the Anthropocene can lead to understandings of socio-spatial transformation as inexorable and transhistorical (Rademacher 2015). In this region, I have demonstrated how these transformations are the product of particular social understandings of gain and loss structured not only by the threat of climate change but also by profoundly unequal political economies. These political economies make some futures thinkable and others unthinkable. Agrarian futures, rendered obsolete by these moral ecologies of the urban Anthropocene, are disrupted through development of aquaculture and associated ecological destruction, or more candidly through imaginaries of forced relocation.

Both materially and epistemically, these moral ecologies of the future for urban livelihoods entail the elision of rural ones. "The story of decline," Tsing writes, "offers no leftovers, no excess, nothing that escapes progress. Progress still controls us even in tales of ruination" (Tsing 2015, 21). These stories from Kolanihat, New Town, and the Sundarbans indicate precisely this rejection of agrarian futures in stories of rural decline in capitalist climate imaginaries. Yet, their stories also suggest the significant political stakes in recognizing the lives, livelihoods, and futures that have been rendered superfluous. And they suggest the political potential in imagining otherwise.

Works Cited

Adnan, Shapan. 2013. "Land Grabs and Primitive Accumulation in Deltaic Bangladesh: Interactions between Neoliberal Globalization, State Interventions, Power Relations and Peasant Resistance." *Journal of Peasant Studies* 40 (1): 87–128.

Afroz, Sharmin, Rob Cramb, and Clemens Grünbühel. 2017. "Exclusion and Counter-Exclusion: The Struggle over Shrimp Farming in a Coastal Village in Bangladesh." *Development and Change* 48 (4): 692–720.

Belton, Ben. 2016. "Shrimp, Prawn and the Political Economy of Social Wellbeing in Rural Bangladesh." *Journal of Rural Studies* 45: 230–42.

Bhattacharyya, Debjani. 2018a. "Discipline and Drain: Settling the Moving Bengal Delta." *Global Environment* 11: 236–57.

Bhattacharyya, Debjani. 2018b. *Empire and Ecology in the Bengal Delta: The Making of Calcutta*. Cambridge: Cambridge University Press.

Brammer, Hugh. 2009. "Climate Refugees: A Rejoinder." *Economic and Political Weekly* 44 (29): 87.

Brammer, Hugh. 2014. "Bangladesh's Dynamic Coastal Regions and Sea-Level Rise." *Climate Risk Management* 1: 51–62.

Cohen, Daniel Aldana. 2016. "The Rationed City: The Politics of Water, Housing and Land Use in Drought-Parched São Paulo." *Public Culture* 28 (2): 261–89.

Collard, Rosemary-Claire, Jessica Dempsey, and Juanita Sundberg. 2015. "A Manifesto for Abundant Futures." *Annals of the Association of American Geographers* 105 (2): 322–30.

Das Gupta, Moushumi. 2017. "Mamata vs Modi Govt: 5 Central Schemes Stonewalled by the West Bengal CM." *Hindustan Times*, April 6, 2017, India. https://www.hindustantimes.com/india-news/mamata-vs-modi-govt-5-central-schemes-stonewalled-by-the-west-bengal-cm/story-DtF3ZJKSvrMOiE8etkd03J.html.

Datta, Anjan Kumar. 1998. *Land and Labour Relations in South-West Bangladesh: Resources, Power and Conflict*. New York: St. Martin's Press.

Datta, Ayona. 2019. "Postcolonial Urban Futures: Imagining and Governing India's Smart Urban Age." *Environment and Planning D: Society and Space* 37 (3): 393–410.
Elliott, Rebecca. 2017. "Who Pays for the Next Wave? The American Welfare State and Responsibility for Flood Risk." *Politics & Society* 45 (3): 415–40.
Elliott, Rebecca. 2019. "'Scarier than Another Storm': Values at Risk in the Mapping and Insuring of US Floodplains." *British Journal of Sociology* 70 (3): 1067–90.
Farbotko, Carol. 2010. "Wishful Sinking: Disappearing Islands, Climate Refugees and Cosmopolitan Experimentation." *Asia Pacific Viewpoint* 51 (1): 47–60.
Foucault, Michel. 2003. *'Society Must be Defended': Lectures at the College de France 1975-1976*. New York: Picador.
Ghosh, Nilanjan, Anamitra Anurag Danda, Jayanta Bandyopadhyay, and Sugata Hazra. 2016. *Away from the Devil and the Deep Blue Sea: Planned Retreat and Ecosystem Regeneration as Adaptation to Climate Change*. New Delhi: WWF-India.
Ghoshal, Aniruddha. 2016. "Green City Mission: Rs 50 Lakh Each for All 125 Municipalities in West Bengal." *The Indian Express*, December 21, 2016. http://indianexpress.com/article/india/green-city-mission-rs-50-lakh-each-for-all-125-municipalities-in-west-bengal-mamata-banerjee-4438068/.
Goh, Kian. 2017. "Terrains of Contestation: The Politics of Designing Urban Adaptation." In *Perspecta 50: Urban Divides The Yale Architectural Journal*, edited by Meghan McAllister and Mahdi Sabbagh, 63–74. Cambridge, MA: MIT Press.
Goh, Kian. 2020. "Flows in formation: The Global-Urban Networks of Climate Change Adaptation." *Urban Studies* 57 (11): 2222–40. https://doi.org/10.1177/0042098018807306.
Haraway, Donna. 2016. *Staying with the Trouble: Making Kin in the Chthulucene*. Durham, NC: Duke University Press.
Iqbal, Iftekhar. 2010. *The Bengal Delta: Ecology, State and Social Change, 1840–1943*. New York: Palgrave Macmillan.
Jalais, Annu. 2010. *Forest of Tigers: People, Politics and Environment in the Sundarbans*. London: Routledge.
Jazeel, Tariq. 2018. "Urban Theory with an Outside." *Environment and Planning D: Society and Space* 36 (3): 405–19.
Knuth, Sarah. 2017. "Green Devaluation: Distruption, Divestment, and Decommodification for a Green Economy." *Capitalism Nature Socialism* 28 (1): 98–117.
Koslov, Liz. 2016. "The Case for Retreat." *Public Culture* 28 (2): 359–87.
Koslov, Liz. 2019. "Avoiding Climate Change: 'Agnostic Adaptation' and the Politics of Public Silence." *Annals of the American Association of Geographers* 109 (2): 568–80.
Li, Tania Murray. 2010. "To Make Live or Let Die? Rural Dispossession and the Protection of Surplus Populations." *Antipode* 41 (S1): 66–93.
Lowry, Kem, Roland Fuchs, and Keith Bettinger. 2015. *Urban Climate Change Adaptation and Resilience: A Training Manual*. Honolulu, HI: USAID.
Marino, Elizabeth. 2018. "Adaptation Privilege and Voluntary Buyouts: Perspectives on Ethnocentrism in Sea Level Rise Relocation and Retreat Policies in the US." *Global Environmental Change* 49: 10–13.
Norwegian Refugee Council. 2015. *Community Resilience and Disaster-Related Displacement in South Asia*. Oslo, Norway: Norwegian Refugee Council.
Padwe, Jonathan. 2019. "Review: The Mushroom at the End of the World: On the Possibility of Life in Capitalist Ruins." *The Journal of Peasant Studies* 46 (2): 433–37.

Paprocki, Kasia. 2016. "'Selling Our Own Skin:' Social Dispossession through Microcredit in Rural Bangladesh." *Geoforum* 74: 29–38.

Paprocki, Kasia. 2018. "Threatening Dystopias: Development and Adaptation Regimes in Bangladesh." *Annals of the Association of American Geographers* 108 (4): 955–73.

Paprocki, Kasia. 2019. "All That Is Solid Melts into the Bay: Anticipatory Ruination and Climate Change Adaptation." *Antipode* 51 (1): 295–315.

Paprocki, Kasia. 2020. "The Climate Change of Your Desires: Climate Migration and Imaginaries of Urban and Rural Climate Futures." *Environment and Planning D: Society and Space* 38 (2): 248–66.

Paprocki, Kasia, and Jason Cons. 2014. "Life in a Shrimp Zone: Aqua- and Other Cultures of Bangladesh's Coastal Landscape." *Journal of Peasant Studies* 41 (6): 1109–30.

Patel, Toral. 2015. "Asia's Cities Poised to Lead in Climate Change Adaptation." *InAsia: Weekly Insights and Analysis*, January 28, 2015. https://asiafoundation.org/2015/01/28/asias-cities-poised-to-lead-in-climate-change-adaptation/.

Podesta, John, and Peter Ogden. 2007. "The Security Implications of Climate Change." *The Washington Quarterly* 31 (1): 115–38.

Pulido, Laura. 2018. "Racism and the Anthropocene." In *Future Remains: A Cabinet of Curiosities for the Anthropocene*, edited by Gregg Mitman, Marco Armiero, and Robert S. Emmett, 116–28. Chicago: University of Chicago Press.

Rademacher, Anne. 2015. "Urban Political Ecology." *Annual Review of Anthropology* 44: 137–52.

Rademacher, Anne. 2018. *Building Green: Environmental Architects and the Struggle for Sustainability in Mumbai*. Oakland: University of California Press.

Rademacher, Anne, and K. Sivaramakrishnan. 2013. "Introduction: Ecologies of Urbanism in India." In *Ecologies of Urbanism in India: Metropolitan Civility and Sustainability*, edited by Anne Rademacher and K. Sivaramakrishnan, 1–42. Hong Kong: Hong Kong University Press.

Reed, Sarah Orleans, Richard Friend, Jim Jarvie, Justin Henceroth, Pakamas Thinphanga, Dilip Singh, Phong Tran, and Ratri Sutarto. 2015. "Resilience Projects as Experiments: Implementing Climate Change Resilience in Asian Cities." *Climate and Development* 7 (5): 469–80.

Roy, Ananya. 2003. *City Requiem, Calcutta: Gender and the Politics of Poverty*. Minneapolis: University of Minnesota Press.

Roy, Ananya. 2016. "When Is Asia?" *The Professional Geographer* 68 (2): 313–21.

Samaddar, Ranabir. 1999. *The Marginal Nation: Transborder Migration from Bangladesh to West Bengal*. New Delhi: Sage.

Shamsuddoha, Md, SM Munjurul Hannan Khan, Sajid Raihan, and Tanjir Hossain. 2012. *Displacement and Migration from Climate Hot-Spots in Bangladesh: Causes and Consequences*. Dhaka, Bangladesh: ActionAid Bangladesh.

Szczepanski, Marcin, Frank Sedlar, and Jenny Shalant. 2018. "Bangladesh: A Country Underwater, a Culture on the Move." In *onEarth*: Natural Resources Defense Council.

Tanner, Thomas, Tom Mitchell, Emily Polack, and Bruce Guenther. 2009. *Urban Governance for Adaptation: Assessing Climate Change Resilience in Ten Asian Cities*. Brighton: IDS: IDS Working Paper 315.

Tsing, Anna. 2012. "Unruly Edges: Mushrooms as Companion Species." *Environmental Humanities* 1: 141–54.

Tsing, Anna Lowenhaupt. 2015. *The Mushroom at the End of the World*. Princeton, NJ: Princeton University Press.

Tyler, Stephen, and Marcus Moench. 2012. "A Framework for Urban Climate Resilience." *Climate and Development* 4 (4): 311–26.

UN-Habitat, and UN ESCAP. 2014. *Pro-Poor Urban Climate Resilience in Asia and the Pacific*. Nairobi: United Nations.

Wachsmuth, David, Daniel Aldana Cohen, and Hillary Angelo. 2016. "Expand the Frontiers of Urban Sustainability." *Nature* 536: 391–93.

Wolford, Wendy. 2005. "Agrarian Moral Economies and Neoliberalism in Brazil: Competing Worldviews and the State in the Struggle for Land." *Environment and Planning A* 37 (2): 241–61.

Zeiderman, Austin. 2016. *Endangered City: The Politics of Security and Risk in Bogotá*. Durham, NC: Duke University Press.

4

The Singapore "Garden City": The Death and Life of Nature in an Asian City

Annu Jalais

In 2019, Singapore fêted its bicentennial "founding" in 1819, and in 2020, fifty-five years as an independent nation. Over the last fifty-five years, this city-state has become increasingly urbanized and wealthy[1] and now has one of the world's densest populations at nearly 8,000 people per square kilometer.[2] It has essentially very few farmlands left. The main primeval vegetation type of Singapore is lowland evergreen rain forest, which originally occupied about 82 percent of the land area, with mangrove and freshwater swamp forests constituting the remainder (Corlett 1991). With the intensification of land development, primary and tall secondary forests dwindled to about 1,700 hectares while built-up areas accounted for about 50 percent of the land area in the late 1980s; this has remained more or less the same to this day (Yee et al. 2011). What remains incredible is how alongside these built-up areas, attention is paid to keeping the place green, which in the context of Singapore is literal; in that this "green" space signifies an assortment of trees and grassy patches along highways, urban well-maintained parks and golf courses.[3] Singapore, the "Garden City," has been a longstanding trope used by government officials to promote tourism and actively entice the world's cosmopolitan well-heeled and moneyed to come to work or settle in the small cityscape. This "green space" sales pitch has also been used to attract tourists. It has been so successful that the Singapore Tourist Promotion Board (STPB), as early as 1987, awarded its

---

1. "Between 1965 and 2013, the city-state achieved a 1,356 percent in real GDP (gross domestic product) per capita growth compared to 146 percent for the world and 96 percent for the United States for the same period" (Savage 2019, 2).
2. According to the statistics portal Statista, the 2016 population density of Singapore was 7,909 people per square kilometer (https://www.statista.com/statistics/778525/singapore-population-density/). This makes Singapore the third most densely populated country in the world (though not one of the most densely populated cities in the world). Of Singapore's 5.5 million population in 2015, 69 percent or 3.8 million are Singaporean citizens and 31 percent or 2.16 million are permanent residents (530,000) and foreigners (1.63 million) residing in the city-state. Thus, close to one-third of Singapore's population comprises foreigners (Savage 2019, 2).
3. "The Building and Construction Authority hopes to have, by 2030, 80 percent of all buildings certified under the 'green mark' scheme. Since its inception in 2005, 2,500 buildings are 'green' in 2015, covering about 29 percent of Singapore's total gross floor area (GFA)" (Savage 2019, 6).

most prestigious tourism award to the Parks and Recreation Department (PRD)—the department in charge of greening Singapore. This greenery was soon touted as important not just to make Singapore a world-class city to attract tourists and foreign investors but also "to lift up" the "spirits" of Singaporeans (K. Y. Lee 1995). "Nature" in Singapore is seen as a resource "that can be shaped to economic and national development objectives" (Yuen 1996, 968). Indeed, what strikes visitors to this garden city is not just its cleanliness and orderliness but also its greenery. Many of Singapore's roads are dotted with beautiful non-native rain trees and bougainvillea. The city is home to one of the best zoos in the world: it has an impressive bird park, there are numerous public parks, and now it offers a river safari with fish species from all over the world. "Nature" in Singapore is a matter of complete governmental control—a resource both in terms of economics to attract foreigners and their investments as well as culture to lift the spirits of home-grown Singaporeans.

In their 2013 book *Ecologies of Urbanism in India*, Rademacher and Sivaramakrishnan propose addressing "the interface between environmental change and urban transformation" when making sense of specific Asian cities (2013, 9), thus bringing in the much-needed cultural dimension to studies on cities. The analytic of urban ecologies, they suggestively argue, "assumes the presence of multiple, simultaneous and overlapping representations of the urban nature-urban culture interface" (2013, 11). In their pioneering 2017 book *Places of Nature in Ecologies of Urbanism*, they went on to say that "unlike a singular ecology that might suggest a unified experience of urban nature" they would like to "identify the multiple forms of nature—in biophysical, cultural, and political terms—that have discernable impact on power relations and human social action" (2017, 3–4). In other words, they stress the importance of the fact that urban ecologies are plural and that acknowledging human social action in relation to these ecologies is important. For this volume, the two authors have urged us to reimagine Asian cities particularly in relation to ecological collapse, death, and destruction. In their concept note they asked us to "rethink the mosaics of urban space" in relation to the themes of life and death (Rademacher and Sivaramakrishnan, May 2017 concept note for the *Death and Life of Nature in Asian Cities* workshop). In what ways, the authors ask, do Asian cities balance entities such as parks, walkways, gardens, and playgrounds and make sense of them in relation to both the actuality as well as the looming threat of the collapse of climate? The key issues they raise, and arguments they advance, are that the biophysical conditions of urban Asian cities and their histories matter, including the colonial legacies that have been woven into the urban landscape. They suggest that to understand cities, especially Asian cities, one has to start "with processes, not borders" (Rademacher and Sivaramakrishnan,[4] this volume, Introduction).

This chapter will argue that in the context of the city-state of Singapore, to be able to make sense of "the interface between environmental change and urban

---

4. See Rademacher and Sivaramakrishnan, "Introduction: Urban Nature Brought to Life in an Age of Loss," in this volume.

transformation" as Rademacher and Sivaramakrishnan ask above (2013, 9), one needs to look at how Singapore's urban nature has been made to navigate through its urban culture and what the city's transformation has meant for its citizens. In other words, what *nature* might mean in this ecology of urban predictability that is Singapore has to do with a certain understanding of the urban culture of Singapore's near-total urban population. This cultivated culture of urbanity, one could argue, goes hand in hand with the ideology of "land scarcity," something that the government promoted since Singapore's status as a sovereign state in 1965. The ruling party of Singapore, the People's Action Party (PAP) government, increased its ownership over land via the Land Acquisition Act in 1966. From the government owning roughly one-third in the 1950s, the PAP government today owns 90 percent of the land; it has justified this land grab on the grounds that "the larger interest of the community must take precedence over the rights of the individual" (Ngiam 2007). The rhetoric of "smallness" and "urban," which was crucial to the postcolonial developmental state's intensified internal territorialisation (Bryant 1998, 36), has not really changed today, especially when Singapore is still referred to as "the little red dot" by its citizens. Indeed, Singapore's public and academic discourse on its own identity playing along the themes of *small* and *urban* perpetuates a political discourse of "survival" and "land scarcity" even after "relative affluence has been achieved" (Goh 2001, 25). Now nearly twenty years after Goh's argument, the discourse might have somewhat changed with the beautiful nature reserves like the Bukit Timah Nature Reserve, the Central Catchment Nature Reserve, and the Sungei Buloh Wetlands gaining traction; however, the impetus for the eradication of natural landscapes by developmental projects is still seen as necessary by the state, and the old Chinese graveyard Bukit Brown Cemetery's significant biodiversity has come under considerable threat, for example, with the government recently allowing a multilane highway through it. As a result, green spaces including both nature areas and gardened parks shrunk from 1.96 hectare per 1,000 persons in 1978 to 0.61 hectare in 1993. Even though in the past decade there has been talk by the Urban Redevelopment Authority (URA) of increasing the greenery of the city to 0.8 hectare per 1,000 population, this has not happened; the government promises to do so by 2030.[5]

Interestingly, however, what has also simultaneously occurred is a kind of perennial conflict between the idea of "development" and the idea of "culture" over nature between government and citizens as revealed in both groups' quest to use nature to reclaim a certain Singaporean cultural identity. Goh argues that the dominant developmental ideologies encompassing governmental discourses were "never completely hegemonic," because the greening of urbanscapes underscored "a fragment of nature in everyday consciousness" and has served "as reminder to the population of another reality far more complex, rich and mysterious than the

---

5. "Designing Our City: Planning for a Sustainable Singapore," https://www.ura.gov.sg/services/download_file.aspx?f=%7B7DFC7DB9-335D-4A12-A072-9C3257269988%7D. Page 17. Accessed October 15, 2019.

urban" (Goh 2001, 14). But I believe that this idea of nature in the context of the Singaporean public also allows a counterargument to governmental notions of nature.

I am going to explore discussions around what nature might represent through a mix of historical and ethnographical methods to highlight internal contradictions about how life or death was viewed in relation to the greening of the city, farms, fishing, and the culling of certain nonhumans—all understood under the umbrella term of "nature"—in this city-scape. Questions discussed in both large and small student groups in the classes I taught while based at the National University of Singapore revolved around what nature might mean for the young in the context of Singapore's urbanity. What was interesting was how these conversations led to reflections on the importance of having a clean and orderly city,[6] and this included discussions on what "live nature" and "dead nature" might mean. Both online and offline, despite an avowedly modern and clinical approach to their cityscape, Singaporeans seemed to be decidedly divided when discussing nature and by extension the place of the "non-human"[7] in the city. My field notes are based on conversations with students who were taking my "Beasts, People, Wild Environments" course and about a dozen citizens with whom I had random conversations in places like supermarkets, food courts, farms, and social media. As of June 2017, the ethnic composition of Singapore is 74.3 percent Chinese, 13.4 percent Malays, 9 percent Indians, and 3.2 percent others, and I would say this reflected the ethnicity of the 2018 class of eighty-nine students taking "Beasts, People, Wild Environments." I consider the kinds of ethical debates that arose in contemplating the place of nature and nonhumans in Singapore in conditions where both greening and infrastructural efficiency are state-sponsored goals in a context of both urban development and urban beautification. Questions around what was meant by the "life" or "death" of nature and nonhumans soon became the emerging themes of these conversations.

As Rademacher and Sivaramakrishnan argue in their introduction to this workshop, *Ecologies and Urbanism in Asia III: Death and Life of Nature in Asian*

---

6. I am not trying to argue it is not. There have been both similar culling as well as spontaneous culling in many parts of the world, such as the massacre of cats by apprentice printers in mid-eighteenth-century France as described by Darnton (1984) or the more recent culling of crows in Yemen's Aden, and that of pigeons in most European and many northern American cities.

7. Following anthropologists such as Viveiros de Castro (1996; 2004), Descola (1996; 2008; 2013) and Bird-David (1999), who have studied the intimate relations humans from various cultures have shared with animals and described how certain cultures' perceptions of "animals" do not fit the Western essentializing dichotomy between the two, I prefer using the term *nonhuman* for animals. I believe this term allows me to include understandings or "ontologies" of "animals" that are not necessarily steeped in the Western nature/society divide, thus allowing for a greater interweaving of an ensemble of socio-animal-natural relations; these have been explained and/or developed more independently by scholars such as Viveiros de Castro (2004), Latour (2004a, 2004b, 2009), and Stengers (2005) and have been at the heart of a few detailed anthropological studies between human and nonhuman animals, such as those of Jalais (2008, 2010, 2018), Willerslev (2007), Govindrajan (2018), Sangma (2016), Sur (2019), and others. They have more recently been at the heart of studies on ethics and social life in South Asia (Sivaramakrishnan 2015) and it offers a broader interpretation than what the term "animal" (a term deeply ensconced within the Judaeo-Christian frame) might mean.

Cities, "in urban studies, the growth of cities is often narrated as a story of planning and disruption, the displacement of non-human life, or the depletion of human and natural resources to concentrate industry, commerce, and government."[8] In this workshop, the authors addressed how the environmental dimensions of cities and urban life are also narrated in terms of death and life. This chapter explores the "simultaneously biophysical and cultural processes through which the death and life of nature are experienced" that Rademacher and Sivaramakrishnan discuss in the workshop in relation to the Asian city of Singapore, which is, despite its projection of controlled equanimity, actually in the throes of a climate change predicament. A study of Asian ecologies of urbanism compels the following questions in the context of Singapore: What, precisely, do citizens allow into their city under the rubric of urban nature? How is nature discerned, defined, experienced, and aspired to in Singapore? What I will attempt to demonstrate is how much of these conversations have to do with certain aspirations for a life not just of development but also of community where individuals come together to form local groups based on an ethos of a shared conception of nature and nonhumans. Around the recurring themes of "life" and "death" that organized these conversations, much was said about what it means to be Singaporean today. It is important to note in this context that both in colonial as well as in postcolonial times, as argued by those who have worked on the environment and the human/nonhuman interface in Singapore (Neo 2012, 2016; Neo and Ngiam 2014; Barnard 2016, 2019; Barnard and Emmanuel 2014; Newman 2014; Chan 2016; Chee 2017; Strand 2018; Corlett 1991, 1992; Tham 2019; Myers 2015), the massive control over "nature" and "non-humans" has defined the history of Singapore from its beginnings and decided who was with the state versus who was against it.

Thinking Life: "Organized" Nature versus "Rebellious" Nature

How Singapore became known as the "Garden City" is best understood through Lee Kuan Yew's speech on October 20, 1995, thirty years after Singapore came into existence:

> Even in the 1960s, when the government had to grapple with grave problems of unemployment, lack of housing, health and education, I pushed for the planting of trees and shrubs. I have always believed that a blighted urban jungle of concrete destroys the human spirit. We need the greenery of nature to lift up our spirits.

Nature, as I argued earlier, in the contemporary Singaporean context, is filtered through human-designed schemes that offer an ordered state of affairs. The words "environmental protection" in the context of Singapore were until the 1990s best understood by state mobilization of the population for anti-litter, resource

---

8. https://www.hkihss.hku.hk/events/ecologies2018/index.html, accessed May 15, 2018.

conservation (water and electricity), and anti-pollution campaigns; these were the mainstay of environmental consciousness among young Singaporeans (Goh 2001, 14). It was a consciousness that was defined by a certain kowtowing to governmental forms of heavy-handed decisions seen as necessary in the face of an assumed (a) land scarcity, and (b) the unquestioned objective rationality of the expert administrator discourse (Goh 2001, 13). But nature has also been the playing field for those who want(ed) to rebel against the government. In other words, politics in Singapore has also often revolved around citizens taking up conservation issues or engaging in subtle ways of push-back against nonhuman culling to challenge governmental decisions. As one in six Singaporeans is a millionaire, the struggle over nature was not really about livelihood but about "quality of life" (Goh 2001, 10), and it still is. I am going to discuss three human-designed schemes to address the internal contradictions nature in Singapore entails: (a) farming and agri-tainment, (b) fishing, and (c) the culling of unwanted (and nonhuman) crows and cats to show how nature is used by the government as well as by various public groups to challenge the government. On the one hand there is the extremely ordered state of affairs that is nature in Singapore from a governmental point of view; on the other hand, nature also (re) presents a space that citizens feel they need to reclaim as their own. It is this tension that is at the heart of urban nature in Singapore.

The "Agri-tainment" Farms of Kranji

What has become important since the 2000s is a new form of "gentleman" farming and something perhaps uniquely Singaporean called "agri-tainment." Over a single weekend in 2005, nearly 10,000 Singaporeans visited the Kranji Countryside Association's (KCA) farms[9] in Singapore's northwest countryside (Tan Wei Xian 2010). The KCA had come up with the term "agri-tainment"—visitors going on farm visits (via chartered, air-conditioned buses), dining on farm-produced, organically grown vegetables (many of which are flown in), and for the more adventurous, even indulging in "farm-stays," which really meant watching "farm-work" (but rarely entailed actually doing farm work). The term "agri-tainment" was soon incorporated by the Singapore Land Authority, the Urban Redevelopment Authority, as well as the Singapore Tourism Board; more recently, it has been used by the Ministry of National Development, the National Parks Board, and the Agri-Food and Veterinary Authority—all government agencies keen to cater to a Singaporean nostalgia for old times and a certain "authenticity." These government agencies were also extremely interested in cashing in on agri-tourism and to highlight to its well-heeled and informed population that Singapore was working toward "sustainability" and that it was more than just a city-state. What is important to know is that before

---

9. This association comprised nine farms and a pottery-kiln.

2005 farmers in Singapore had been fighting a losing battle. Farms and kampungs[10] had all but vanished under the dual exigencies of urbanization and development and government restructuring of the farming sector, especially between 1965 when there were about 25,000 farms to the 1980s when practically no farms remained (Henderson 2009, 265). With the passing of the Land Acquisition Act in 1966, as I mentioned earlier, the state essentially took possession of land and tender, for agricultural land was leased out on a twenty-year basis. Many farmers had to resettle or abandon agriculture altogether. Farms were seen to be at odds with "national developmental goals" as they were seen as "polluting" (pig and chicken farming) and "backward" (because they involved manual labor).[11]

There might have been nostalgia for the old life of the kampungs in the 1990s (Chua 1997), but the kampung no longer exists in today's Singapore and rurality is therefore a thing of the past (Tan Wei Xian 2010). But interestingly, as Tan Wei Xian argues, when trying to bring the rural to the present, "the Singaporean rural ends up referring to some sort of idealised version of a British countryside" (2010, 17). In other words, in the struggle to define the image of Singapore, the anglicized, Western-educated leaders of Singapore believe that "modernization and national success emerged from urban industrial growth, as in the American and European historical experiences" (Chan 2016, 306). Therefore, the image of farming in Singapore is based on a Weberian modern disenchantment with what rurality stands for. The Singaporean Kranji farm owners, for example, seem keen to be seen as belonging to a culture firmly ensconced within a Western paradigm. The Facebook page for Gardenasia offers tips on how to make chamomile tea—a plant that does not grow in Singapore. In a similar vein, Sustenir Agriculture offers strawberries and arugula but not native equivalents such as the jamun or black plum, the Bengal currant, or the bitter gourd. While the Western world may have discovered the moringa, the bitter gourd, and the black plum as the newest health foods (all plants that thrive in the Singaporean soil and climate), Singapore's trendy farms do not offer these on their menu. However, farmers highlight the "Asian" values of hospitality and feeding to entice farm-stays, but the hierarchies of labor remain very much imbricated in Asian economies of hierarchy with the Bangladeshi or Indian migrant doing all the tilling while the Singaporean farmer sits in his air-conditioned farmhouse drawing Excel sheets.

The farms are presented to the public as both fun (following Bryman's Disneyization theory; see Bryman 1999, 40) as well as trendy. An article in *The Straits Times* highlighted a tech-savvy bunch of young farmers under thirty; the article, however, spoke more about the older farmers (Yeo 2010). As Bjorn Low, a

---

10. Kampung is today used to refer to "village" and the pre-HDB settlements in which Singaporeans lived. HDB stands for "Housing and Development Board" and is the agency which regulates the excellent public housing system for Singaporean citizens.
11. Today, all farms come under the jurisdiction of the Singapore Food Agency (SFA) and still manage to produce 13 percent of all the vegetables, 9 percent of the fish, and 24 percent of the eggs consumed in Singapore (SFA 2019).

thirty-nine-year-old cofounder of Edible Garden City, says, "Nature is a very good teacher. It teaches the principle of balance. When I started, I got agitated with the pests everywhere, such as caterpillars and aphids. Now, I realize it's part of nature" (V. Lee 2016). As Nilsson argues, one of the main drivers of agri-tourism is the romanticization of the farmer's lifestyle (2002, 8), and as Tan Wei Xian pertinently develops, some of the owners of these farms, people such as Ivy Singh-Lim for example, the owner of Bollywood Veggies, or Leon Hay from the Hay Goat Dairies, enjoy celebrity status. But many of these "farms" are not farms in the strict sense of the term because, apart from a few fish ponds, they do not keep any animals; this is the case for most of the visitable farms of Singapore. Neither do they own much land, and only part of the food offered in their restaurants comes from what looks more like a large kitchen garden than an actual farm. Many of the farms are designed for visitors, their main attraction being the organic restaurant, even though much of the food is sourced from Australia, New Zealand, or even Europe. One can hold events or even indulge in farm-stays at either Bollywood Veggies or Gardenasia. When visitors book a tour for the Hay Dairies, they go less to help feed the goats, clear their pens, or milk them but more to "learn" about them in a theoretical way. As an article in *The Straits Times* describes these second-generation farmers, they're "young, without a tan and business savvy" (Tan 2016) and so are those visiting them.

This is why, in the context of Singapore, many of these "new" KCA farmers do not necessarily come from a farming background that had, like elsewhere, occupied the same piece of land for generations. What is interesting in relation to these modern farms is how their very existence makes us rethink both the urban as well as the rural, as well as notions of "East" and "West."[12] Occupying 1,500 hectares, there are seemingly around two hundred farms in existence today and what is stressed about these farms is their "scientific" nature and how joining the farming industry is about "research"—one that is understood to be text based rather than practice based. Nowhere does the toil of farming ever seem to appear: the early hours, the back-breaking work, the inability to take a break, and the general drudgery of farm life. Most articles on farming in Singapore on Google, for example, highlight how the city-state is leading technologically. However, most plants still need earthworm-filled soil (even though hydroponic methods are becoming increasingly popular at these farms) and farm animals need to defecate and their enclosures cleaned, but these things rarely figure on any of these farms' promotional sites and videos.

The other aspect around "invisibility" concerns the migrant or foreign workers—the group of people who actually till the soil, plant the seeds, and clean the pens and coops. This is because these farms are portrayed as zones of "learning"

---

12. Let us be clear here that this reintroduction of "nature" through these farms named Bollywood Veggies, Gardenasia, and Hay Dairies Goat Farm are not a return to Kampung life but are instead based on the romanticising of the rural (via the use of the language for the British countryside), especially in contrast to the city. This contrast is a particularly modern notion that has European origins (Bonner 1997).

for children and for "recreation" for adults, where one might be able to indulge in good organic food. They are thus visited mainly by trendy Singaporeans, expats, and visiting Western tourists keen to show their grandchildren how milk and eggs materialize. Very few are nostalgic of a Singaporean kampung past, or a past where people were closer to nature and nonhumans. Despite the tech aspect, what was interesting to hear from a couple of "farmers" was how they wanted to keep things "simple"—a sort of alternative to the hyper-urban "nature" of gardens by the bay. As one of the Bangladeshi cooks at the Poison Ivy restaurant of Bollywood Veggies farms said, "what the super trees are at gardens-by-the-bay, the foreign lifeless super tomatoes are on your plate, both will dazzle you but can your soul ever be replenished by all this fakeness, your stomach really fed?" An Indian migrant worker said, "looking at this kind of green is food for human eyes, not like the fruitless greenery you see elsewhere in town." What is not spoken about are often the deplorable conditions in which migrants work—something that has come to everyone's attention with the explosion of COVID-19 infections amongst the migrant workers of Singapore. A pregnant earthy productive "nature" contrasts to the beautiful albeit "dead" nature of the city, and a farmer sits in an air-conditioned room while a migrant or someone considered a social inferior tills the land and does the heavy-lifting work for him. What was also noticed is the important government investments in these farms (S$63 million invested by the Ministry of National Development in 2014 and nearly S$40 million in 2020 by the Singapore Food Agency (SFA) as part of the country's recent 30X30 Express, which aims to increase Singapore's local food production to meet 30 percent of its demand by 2030), yet the relative disinterest in migrants' working conditions by the majority of the Singaporean citizenry who see these farms more as trendy tourist attraction spots than nostalgic kampong sites.

So "nature" here is a "foreign" farm the government wants to develop. It remains a zone that most Singaporeans are not really interested in visiting or engaging with as they don't really see any part of their identity linked to these farms. For decades the population was fed on dreams of a concrete paradise, an air-conditioned city where their feet would wear shoes and never get muddy again and backs would never get a tan or the ache inflicted by the back-breaking tough life of farming. Yet the government persists in funding these "farms" and the gentlemen (and gentlewomen) farmers that run them. "It gives the Government a good name, you see," explained one of my students. "It gets PR points from other countries for trying to go organic, just like in the West; we are hip."

Dead Life, Dirty Food: Fishing and Foraging

Let us now turn to nature as food, because food really offers a rhetoric that touches upon both life and death. Food offers an analytical lens, as have argued Kong and Sinha—two scholars firmly based in Singapore—which "enables society, culture, economy and polity to be scrutinised and theorised" (2016, 2). As these authors have argued in the introductory chapter to their edited book on food in postcolonial Singapore, there is a certain disjuncture between the producers of food and the consumers of food. Even though in the last couple of decades people have started being more aware and conscious about the source of their food, there is no talk about it beyond it being pesticide-free; "Please don't buy any food coming from China, they did not hesitate adulterating baby milk, who knows what else they poison," an elderly Singaporean told me while we were both shopping at Sheng Siong—the local alternative to Cold Storage, the latter seen as a supermarket especially for "Ang Mohs" or white expatriates (personal communication). Others, essentially expatriate Indians, told me they only trust the vegetables and fish found at Mustafa in Little India because these "were from India" (personal communication). And yet Singapore has probably the most stringent laws in Asia when it comes to food control. What was interesting for me to witness was how food was understood—whether it was seen as adulterated or pure was often seen as depending on its provenance. No conversations either in the public or the private sphere were ever entertained about the ethics of flying food from thousands of miles away, of eating fruits and vegetables outside seasons, or about providing fair wages to workers in Singapore.

Apart from the concern with pesticides, the other aspect of whether something was "eatable" depended on whether it was viewed as wild or not. If most Singaporeans have no problem in seeing certain animals as food, they definitely winced at the thought that foraged fruits, vegetables, greens, fish, or chicken from common spaces such as forested areas or bodies of water such as reservoirs, lakes, or the sea might be "eatable." Recently, especially since the Year of the Rooster (2017), Singaporeans have been surprised to find many non-domesticated roosters, hens, and chicks out in the open, clucking away without a care in the world. And not just in places like the botanical gardens but also in the green patches around Singapore's public housing popularly known as HDBs or even on university campuses. These chickens have kept increasing in number since 2017 as they have very few natural predators. People have been complaining on social media about them as they tend to defecate, like any fowl, all over their pecking area. I asked an elderly Chinese Singaporean if anyone caught and ate these chicken. "No, no, how could we, they're not domesticated!" he replied horrified. "But they're chicken," I probed. "No, they look nice, let them roam lah," he said, smiling. When I asked a young lady why many people complained about them but nobody caught them, she said "they're outside creatures, who knows what they eat?" The other aspect why most people would never think of eating them is because they are seen as "wild" and therefore the possible host-receptors of diseases—this was a recurring point in conversations. Singaporeans unfailingly pointed out that SARS, COVID-19, Ebola, or MERS had been triggered by the consumption of "wild animals."

Similarly, some of my young students who were fishing aficionados often got thrills fishing illegally in the MacRitchie reservoir at dawn. They said, however, that they always returned the fish they had caught due not so much out of a concern for the fish or because "this was just a sport" but because of "a concern about hygiene." Most of the young anglers did so in some of Singapore's purest water reservoirs, like the MacRitchie lake, which provides part of the fresh water the population drinks, and yet the naturally fished organic fish was considered by them as dirty and not fit to be eaten. They also looked down on those less wealthy, often poorer Singaporeans, who legally fished fish from the sea front for personal consumption. They said that food was what was found at the supermarket and not swimming in the sea or a natural reservoir. This dichotomy between the two spheres was always very clearly maintained. It got me interested in what was seen as "dirty" food consumed by "dirty humans." Nappi's work on the Yeti in early modern China explores the ways in which the Chinese carved out the borders of what is meant to be human, and fascinatingly, a lot of it had to do with eating (2012, 70). The kind of food one eats plays an important role in reinforcing cultural and religious ideas about nonhuman animals and, as in all cultures, "the Chinese use food to mark ethnicity, religious and cultural festivals, rites of passage, family events and social transactions" (Donovan 2004, 95). There are of course many cultural distinctions between the Chinese and Singaporeans of Chinese ancestry, but as Lily Kong and Brenda S. A. Yeoh have argued, the power relations that define and contest the specificities of the Singaporean nation are negotiated through "elements of the cultural landscape, including the landscape and practices of everyday life" (2003, 15). A lot of the elements of this cultural landscape, as my students made me understand, had much to do with what you consider as food, both as a cultural legacy as well as one fitting modern notions of hygiene. Hence, the urge my students had to tell me how on trips to China there is a custom to feed on rare creatures but how they themselves could not partake of that as they found the eating of wild animals or fish completely disgusting. This was not lost on me. It was a way for many Singaporean students of Chinese ancestry to mark themselves out as hygienic modern Singaporeans who refused the wild meat delicacy dishes of both the elite Chinese as well as the poached meat of the poorer people. What was fascinating to note was how "dead" meat or fish found in a supermarket was considered "clean" and "hygienic," whereas wild game or fish fished from one of the reservoirs that provides drinking water to the population was considered dirty and unhygienic.

The State-Culling of "Out of Place" Crows and Cats

Thousands of crows have been culled in Singapore since the mid-1980s, and it is perhaps the only city in Asia where one will rarely see a crow. Similarly, cats are not allowed in HDBs as they are seen by the government as "uncontrollable" and "dirty" as they do not respect the inside/outside boundary. House crows (*Corvus splendens*) are ubiquitous to most Asian cities. Crows are highly adaptable and are often the ones to clean up detritus from open-air dumping grounds in the world's warmer cities and small towns. Some Singaporeans I spoke to had witnessed first-hand crow-culling when looking out through classroom windows as schoolchildren, but practically nobody mentioned the culling as having been distressing; they kept insisting that it was important as it was for the greater good of Singapore and its citizens. Respondents legitimized the culling on the grounds that (a) crows were alien to Singapore, and (b) they were dirty and vectors of diseases. Indeed, these two points were also insisted upon in articles about crows in Singapore. In a piece called "The 'Aliens' in Singapore," published in *Nature Watch*, Lim Kim Seng (member of the Conservation Sub-committee Nature Society), writes:

> The House Crow was introduced during World War II to combat a plague of caterpillars. It is not known whether they managed to contain the plague but what is known is that this common bird has itself become a nuisance. The house crow spread steadily and in 1985 it had successfully colonised the whole of Singapore, including Pulau Ubin. A few crows may carry contagious diseases like avian pox and infect other resident birds. (K. Lim 1998)

In another provocatively titled article on house crows, "Undesirable Aliens," scientists H. C. Lim, Sodhi, Brook, and Soh (2003), based in Singapore, write about three invasive bird species—the house crows, the white-vented myna, and the common myna—as being "biological invasions" that have negative impacts on biodiversity and economics. Undertaking a survey in 2000, these scientists argue that the house crow had increased dramatically since the mid-1980s even though culling had started in 1973 in Singapore. The house crow, referred to in one paper as the Indian house crow (Brook et al. 2003, 808), was seen as more undesirable, because its "constant loud cawing, habit of feeding on rubbish, and its black plumage as a superstitious symbol of bad omens" made it a much more conspicuous pest bird in Singapore than the white-vented myna (Lim et al. 2003, 692). With average populations for the house crow, in 1985, numbering 132,000 individuals, the aim of these scientists was, with the start of the 2000s, to try and exterminate this species by culling 41,000 to 44,000 of them in 2002 and then spend equivalent labor hours each year thereafter for a ten-year period through both shooting and destroying of nests to bring the crow population drastically down (Brook et al. 2003, 813–14). Since arriving in Singapore in 2012, this author has noticed the absence of crows and has seen one occasionally only about half a dozen times in all. Whether native avifauna is thriving due to this culling is another question.

The fact that the crow is supposedly "alien" is a recurring trope and one that is used to justify culling. But in a way, like for the rest of the introduced flora and fauna, crows' "foreignness" is up for debate. As van Dooren highlights, crows have always been like stowaways on ships roaming around the globe since the 1800s or brought intentionally by people in what biologists term "self-introduction," to the extent that they are now found in about twenty countries outside of what is considered to be their native range and are not found to be living separate from humans—that is, the species is now considered to be an "obligate human commensal" (2016,

195). An article published in *The Straits Times* talks about the "'attack" of exotic flora and fauna on Singapore (Vaughn 2010). The article mentions how the definition of "invasive species" needs to be broadened to include "nuisance or pest species such as the African giant land snail, the Javan mynah, the house crow[13] and the American cockroach" (Vaughn 2010). The article further discusses how the urban environment of Singapore, however, also comprises many introduced species. So, if not all "foreign talents" are invasive, it is usually unwanted plants and animals that are termed *invasive* as opposed to those which are welcomed and tolerated and therefore termed *introduced*. Indeed, the beautiful and easily recognizable umbrella-shaped Rain Tree that provides shade along most of Singapore's perfect roads, a tree that could be called Singapore's signature tree, was introduced as recently as 1876 and originally comes from Central and South America. Therefore, "foreign talent" is double-edged; it is either "invasive" and unwanted (such as the house crow) or "introduced" and tolerated (like the laughing thrush) or even feted (like the Rain Tree).

This is where dismissing stray cats by including them in the "foreign" category becomes difficult. Stray cats are very much both local as well as unwanted by the government. Since September 1978, Singapore's HDB, regulates what it calls HDB-approved pets, which include smaller dog breeds, birds, hamsters, and rabbits. Cats, however, have been excluded from the approved list because they are deemed ungovernable. The HDB website states:

> Cats are not allowed in flats. They are generally difficult to contain within the flat. When allowed to roam indiscriminately, they tend to shed fur and defecate or urinate in public areas, and also make caterwauling sounds, which can inconvenience your neighbours.[14]

Indeed, as artist, art critic, and stray cat lover Lucy Davis highlights, a cat does "not follow the grid system of the city. It does not walk on a leash and cannot be trained. It is unpredictable, nocturnal, transgressive" (2011, 194). But even though cats are banned from public housing, they are found in most of them: "illegally kept cats can be found resting around staircase landings, or strolling nonchalantly along the corridors and void decks"[15] (Chee 2017, 1060).

Lilian Chee argues that government policies are against pet cats in HDBs because they transgress between indoors and outdoors, and they are thus perceived as potential vectors of contagion. Therefore, "putting the cat out at night" signals incomplete containment in the home, only a partial domestication, even though cats may be cosmetically modified to fit conceptions of the homely and the domestic. Similarly, the cat flap is a breach in the domestic boundary, and cats bringing mice or birds into the home may be seen as polluters of domestic space. In

---

13. Also referred to as the "Indian" house crow.
14. www.hdb.gov.sg/cs/infoweb/residential/living-in-an-hdb-flat/keeping-pets, accessed October 15, 2019.
15. The "void deck" is the open public space on the ground level of any public housing block.

this respect, pet cats are transgressive, breaking the boundary between nature and culture, between the home and the world. These free-roaming cats exemplify

> the ambiguous private-public boundaries that occupants regard and behave within on the void deck. The void deck's completely porous and open edges are contradictorily matched by its exact mapping of the residential units' structural grid: one gesturing to unregulated access; the other a physical reminder of every unit's fiercely guarded privacy. It is a space that feels familiar and alienating at the same time. This spatial ambiguity gives anonymity to the stealthy cat owners, who are also assuaged that their cats are roaming not far from home. (Chee 2017, 1060)

But perhaps, as Chee (tongue-in-cheek) points out by quoting Davis, what is really the cause for governmental alarm is that the cat "has sex very loudly and uses its sexuality 'irresponsibly,' reproducing out of control" (2017, 1060).

What both Chee and Davis stimulatingly explore is how Singapore presents a rather rich territory to study how nonhumans figure in peoples' minds. Davis underscores how advertisements against "pests," whether cockroaches or the *Aedes* mosquito, are talked about in gendered terms: the male feed on fruit, but the female needs human blood to mature her eggs. So the dengue-causing *Aedes Aegypti* mosquito has morphed, in popular representations, into the South East Asian phantom of the *Pontianak* or the South Asian blood-sucking female vampire called *churail*, explains Davis. She also takes the example of the partially autobiographical novel *Wuya* ("crow" in Mandarin) written by Zhu Ziping (pen-name Jiu Dan), a female author from mainland China who appropriated this moniker to delve into the supposed sexual and money-grabbing underworld of her migrant compatriots to Singapore. These single women, or single mothers, accompanying their children to Singapore, while their husbands remain in China, are derogatively called "crows" out to get permanent residence status by "stealing" Singaporean men. When the novel came out in 2001, it caused quite a controversy,[16] with mainland Chinese in Singapore and China protesting against the offensive portrayal of the mainland Chinese women who supposedly come into the city-state to "steal" the husbands and the benefits of Singaporeans (Davis 2007, 180). Through these metaphors of mosquitoes/*Pontianaks*, cats/deranged lawless single women, and crows/foreign agents who come to Singapore to steal wealth and men, Davis highlights how the portrayal of these nonhumans shapes Singaporean culture, society, and beliefs. This "gendering" of cockroaches, mosquitoes, crows, and cats, and the parallel representing of unwanted humans in gendered terms, feed into unequal power relations between those who hold power versus those who do not, the state versus "outcast" citizens and unwanted (yet very much needed) "foreign" or "guest" workers. As one of my cat-feeding friends said, "I am a Buddhist and I believe in non-violence and in doing good around me, how can I not feed these hounded creatures?"

---

16. See BBC News World Edition 2001 article for details (http://news.bbc.co.uk/2/hi/entertainment/1475842.stm, accessed October 15, 2018).

When the SARS outbreak occurred in Southeast Asia in 2003, the Singaporean government was seen using this event to start cat culling as well as continue crow culling more aggressively. The governmental language was one that argued that these stray cats and crows could spread the disease. But a small group of ardent animal lovers and activists decided to rebel against the government by hiding cats and feeding them secretly (not so the crows). "When they went after our cats, it felt like they were after us, us single and unmarried and not reproducing and therefore seen as useless by the state," said one of my female middle-aged and unmarried interlocutors. "Stray cats go against the government's ideology," explained some of those I spoke to, "because they cannot be kept on leash" and "they are temperamental and noisy," said another (personal communication). Drawing from her own relationship with the Singapore feline network, Davis (2011) argues that there is an obvious gender bias against the volunteer cat feeders. Many of these "nocturnal cat-feeders" are women who are usually "single, divorced, gay, childless, or with children moved away," thus departing from the normative mother/wife profile indicated for public housing tenancy (Davis 2011, 189). These "cat-women," just like the cats they feed, are not necessarily residents of the same HDB. Davis highlights the figure of the cat and of these cat-feeding women to make a point about the parallels between the conception and the demonization of certain nonhumans and the vilification of a certain group of people—in this case often eccentric, single women, seemingly unattached to child-rearing. Davis makes a critical connection with the graduate mother scheme,[17] where single women were encouraged to marry and reproduce, for "the good of the city" (those with lesser education, on the other hand, were given incentives to go for irreversible birth control) and the cat culling raised issues about how the scheme led to a greater surveillance and control of society by those in power and how this was deeply resented (2007, 186).

Conclusion: Life and Death in the Asian Anthropocene

This "overlapping" of creatures "interbodying with" a section of the population and spaces such as agri-farming "functioning multivalently" has become the new paradigm when studying nature in the Anthropocene. In *Wildlife in the Anthropocene*, Lorimer (2015) argues that contemporary studies on animals allows one to connect science, politics, and popular culture as well as real-world interactions with the natural environment. This is because in the Anthropocene, Lorimer argues, the natural world is hybrid, nonlinear, and multiple. Strangely, or not, this perception of

---

17. The graduate mother scheme of 1983 was a government procreation scheme that gave financial incentives for graduate women to produce more children and similar incentives for low education, low-income families who in turn agreed to not have any more. Two groups of "deviant women" were identified, firstly, graduate single women (read: Chinese) who were not getting married "quickly enough" or were not producing enough children and the uneducated minority working-class (read: Malay and Indian) married women who were causing concern by producing too many purportedly less intelligent, less productive offspring (Heng and Devan, 1992).

the world was already one that many societies shared before joining the community of the "moderns" (as exemplified by Latour's 1993 study). Inspired by Whatmore's influential *Hybrid Geographies* (2002), Lorimer decides in his book to invest "wildlife" with new meaning. In these post-nature times, he argues, "wildlife lives among us" and includes "the feral plants and animals that inhabit urban ecologies" (Lorimer 2015, 7). What excites Lorimer about this wildlife that lives among us is its inherently non-anthropocentric, differentiated, and ever-changing character, one that "flags the degree to which any management decision is a biopolitical act," as it is a managing of life at the level of populations (2015, 33). Lorimer helps us connect the multifaceted case studies in Singapore: the ways in which nature is understood and highlighted in farming and agri-tainment as something undertaken by trendy, tech-savvy youngsters; how wild animals are seen as vectors of disease and hence "dirty food" only fit for the (essentially) poor; and how the culling of crows and cats stands in for a sort of female defiance to the state. What ties up the plant world, the water bodies, and the nonhuman world is how all of these are realms that are both ensconced within the rhetoric of "development" as used by the government as well as taken over by the noncomplying actions and rhetoric of many Singaporeans who focus on ideas of the past, on an anti-consumerist ethos, or on a specific cultural identity. These examples emphasize the entangled relationship between "nature" and a city-state fixated with cleanliness and efficiency. As Clancey has recently argued, "Public health bureaucracies came to play an important supporting role in the creation of this 'landlord state', in which health became imbricated with cleanliness and habitation, all three becoming realms of state responsibility" (2018, 214). An approach to nature fragmented along two citizenries—one obsessed with status and hygiene, the other happy to discreetely forage for fruits, engage in petty poaching, and care for the hidden lives of cats. What is interesting is to see how shared space with plants via farm associations such as KCA as well as shared space with boundary-defying cats and crows have become ways in which some citizens have decided to stand up to what they see as the strong-arm of the government without it necessarily translated toward a demand for greater justice toward migrant workers or care toward poorer or lonely (and often elderly) Singaporeans.

Singapore, in many ways, sees itself as a laboratory that comes up with ways to defy the effects of "nature." It is an island that has largely been growing in geographical size, especially in the last three decades, via the extortion of sand from poorer Southeast Asian countries for purposes of land reclamation. Singapore is also one of the lowest-lying coastal cities that will be in the direct firing line of rising sea levels; though the government has been planning for this eventuality by embarking on a program to raise its coastal areas by around 2 meters above the current sea level to buffer sea-level rises (Grundy-Warr and Savage 2017, 463), there is hardly any discussion in the public sphere about the threat of climate change to the city-state.[18]

---

18. There is, however, an excellent recent article on how the Indian Ocean region needs to come together to address climate change by Savage and Lin (2020).

This is why understanding urbanization using Singapore as a case study forces us to confront the tension the editors of this book want us to grapple with: Singapore, like many Asian cities, and perhaps more than any other Asian city, seems to embody both limitless potential and impending doom all at once and many of these conversations, led by its citizens, revolve around nature and its living and dying.

Acknowledgments

The author would like to thank Professors Sivaramakrishnan and Rademacher for their invitations to the two workshops organized to discuss the themes that have been explored in this chapter. The author would also like to thank Yukiko Tonoike for her infinite patience and her excellent editing skills and Lin Qi Feng for a very critical reading of this chapter and some excellent references. Mistakes, however, remain very much my own.

Works Cited

Barnard, Timothy P. 2016. *Nature's Colony: Empire, Nation, and Environment in the Singapore Botanic Gardens*. Singapore: NUS Press.

Barnard, Timothy P. 2019. *Imperial Creatures: Humans and Other Animals in Colonial Singapore, 1819–1942*. Singapore: NUS Press.

Barnard, Timothy P., and Mark Emmanuel. 2014. "Tigers of Colonial Singapore." In *Nature Contained: Environmental Histories of Singapore*, edited by Timothy P. Barnard, 55–80. Singapore: NUS Press.

BBC News World Edition. 2001. "Chinese Prostitute Book Sparks Outrage." *BBC News World Edition*. August 6, 2001. Accessed January 10, 2018. http://news.bbc.co.uk/2/hi/entertainment/1475842.stm.

Bird-David, Nurit. 1999. "'Animism' Revisited: Personhood, Environment, and Relational Epistemology." *Current Anthropology* 40 (S1): S67–S91.

Bonner, Kieran. 1997. *A Great Place to Raise Kids: Interpretation, Science, and the Urban-Rural Debate*. Montreal: McGill-Queen's University Press.

Brook, Barry W., Navjot S. Sodhi, Malcolm C. K. Soh, and Haw Chuan Lim. 2003. "Abundance and Projected Control of Invasive House Crows in Singapore." *The Journal of Wildlife Management* 67 (4): 808–17.

Bryant, Raymond L. 1998. "Resource Politics in Colonial South-East Asia: A Conceptual Analysis" in *Environmental Challenges in South-East Asia*, edited by Victor T. King, 27–52. Surrey: Curzon Press.

Bryman, Alan. 1999. "The Disneyization of Society." *The Sociological Review* 47: 25–47.

Chan, Ying-kit. 2016. "No Room to Swing a Cat? Animal Treatment and Urban Space in Singapore." *Southeast Asian Studies* 5 (2): 305–29.

Chee, Lilian. 2017. "Keeping Cats, Hoarding Things: Domestic Situations in the Public Spaces of the Singaporean Housing Block." *The Journal of Architecture* 22 (6): 1041–65.

Chua, Beng-Huat. 1997. *Political Legitimacy and Housing: Stakeholding in Singapore*. London: Routledge.

Clancey, Gregory. 2018. "Hygiene in a Landlord State: Health, Cleanliness and Chewing Gum in Late Twentieth Century Singapore." *Science, Technology and Society* 23 (2): 214–33.

Corlett, Richard. 1991. "Vegetation." In *The Biophysical Environment of Singapore*, edited by L. S. Chia, A. Rahman, and D. B. H. Tay, 134–54. Singapore: Singapore University Press.

Corlett, Richard. 1992. "The Ecological Transformation of Singapore 1819–1990." *Journal of Biogeography* 19: 411–20.

Darnton, Robert. 1984. *The Great Cat Massacre and Other Episodes in French Culture History*. New York: Basic Books.

Davis, Lucy. 2007. Notes for a Singapore Bestiary: Gender, Sexuality and Inter-Species Encounters in the City State. FOCAS: *Forum on Contemporary Art & Society* (Vol. 4). Singapore.

Davis, Lucy. 2011. "Zones of Contagion: The Singapore Body Politic and the Body of the Street-Cat." In *Considering Animals: Contemporary Studies in Human-Animal Relations*, edited by Carol Freeman, Elizabeth Leane, and Yvette Watt, 279–302. Farnham, Surrey: Ashgate Publishing.

Descola, Philippe. 1996. "Constructing Natures: Symbolic Ecology and Social Practice." In *Nature and Society: Anthropological Perspectives*, edited by Phillipe Descola and Gíslí Pálsson, 82–102. London: Routledge.

Descola, Philippe. 2008. "Who Owns Nature?" *Books and Ideas*. January 21, 2008. Accessed January 15, 2018. http://www.laviedesidees.fr/Who-owns-nature.html?lang=fr.

Descola, Philippe. 2013. *Beyond Nature and Culture*. Chicago: University of Chicago Press.

Donovan, Deanna G. 2004. "Cultural: Underpinnings of the Wildlife Trade in Southeast Asia." In *Wildlife in Asia: Cultural Perspectives*, edited by John Knight, 88–111. London: Routledge.

Goh, Daniel. 2001. "The Politics of the Environment in Singapore? Lessons from a 'Strange' Case.'" *Asian Journal of Social Science* 29 (1): 9–34.

Govindrajan, Radhika. 2018. *Animal Intimacies: Interspecies Relatedness in India's Central Himalayas*. Chicago: University of Chicago Press.

Grundy-Warr, Carl, and Victor R. Savage. 2017. "Singapore: Sustaining a Global City-State and the Challenges of Environmental Governance in the Twenty-First Century." In *Routledge Handbook of the Environment in Southeast Asia*, edited by Philip Hirsch, 448–69. New York: Routledge.

Henderson, Joan. 2009. "Agro-tourism in Unlikely Destinations: A Study of Singapore." *Managing Leisure* 14 (4): 258–68.

Heng, Geraldine, and Janadas Devan. 1992. "State Fatherhood: The Politics of Nationalism, Sexuality and Race in Singapore." In *Nationalisms and Sexualities*, edited by Andrew Parker, Mary Russo, Doris Sommer, and Patricia Yaeger, 343–64. New York: Routledge.

Jalais, Annu. 2008. "Unmasking the Cosmopolitan Tiger." *Nature and Culture* 3 (1): 25–40.

Jalais, Annu. 2010. *Forest of Tigers: People, Politics and Environment in the Sundarbans*. New Delhi: Routledge.

Jalais, Annu. 2018. "Reworlding the Ancient Chinese Tiger in the Realm of the Asian Anthropocene." *International Communication of Chinese Culture* 5 (May 9): 121–44.

Kong, Lily, and Vineeta Sinha. 2016. *Food, Foodways and Foodscapes: Culture, Community and Consumption in Post-Colonial Singapore*. Singapore: World Scientific Publishing.

Kong, Lily, and Brenda S. A. Yeoh. 2003. *The Politics of Landscape in Singapore: Constructions of 'Nation'*. Syracuse, NY: Syracuse University Press.

Latour, Bruno. 1993. *We Have Never Been Modern*. Translated by C. Porter. Cambridge, MA: Harvard University Press.
Latour, Bruno. 2004a. *The Politics of Nature: How to Bring the Sciences into Democracy*. Cambridge, MA: Harvard University Press.
Latour, Bruno. 2004b. "Whose Cosmos? Which Cosmopolitics? Comments on the Peace Terms of Ulrich Beck." *Common Knowledge* 10 (3): 450–62.
Latour, Bruno. 2009. "Perspectivism: 'Type' or 'Bomb'?" *Anthropology Today* 25 (2): 1–2.
Lee, Kuan Yew. 1995. Speech given at the launch of the national orchid garden. October 20, 1995. Accessed October 15, 2019. https://www.nas.gov.sg/archivesonline/data/pdfdoc/lky19951020.pdf.
Lee, Venessa. 2016. "Life Interview with Bjorn Low: Urban Farming Gave Life Goal to Co-Founder of Edible Garden City." *The Straight Times*. October 3, 2016. Accessed October 15, 2019. https://www.straitstimes.com/lifestyle/a-farmer-in-the-city.
Lim, Haw Chuan, Navjot S. Sodhi, Barry W. Brook, and Malcolm C. K. Soh. 2003. "Undesirable Aliens: Factors Determining the Distribution of Three Invasive Bird Species in Singapore." *Journal of Tropical Ecology* 19 (6): 685–95.
Lim, Kim Seng. 1998. "The 'Aliens' in Singapore." *Nature Watch* 6 (3). Accessed October 15, 2019. http://habitatnews.nus.edu.sg/pub/naturewatch/text/a063b.htm.
Lorimer, Jamie. 2015. *Wildlife in the Anthropocene: Conservation after Nature*. Minneapolis: University of Minnesota Press.
Myers, Natasha. 2015. "Edenic Apocalypse: Singapore's End-of-Time Botanical Tourism." In *Art in the Anthropocene: Encounters Among Aesthetics, Politics, Environments and Epistemologies*, edited by Heather Davis and Etienne Turpin, 31–42. London: Open Humanities Press.
Nappi, Carla. 2012. "On Yeti and Being Just: Carving the Borders of Humanity in Early Modern China." In *Animals and the Human Imagination: A Companion to Animal Studies*, edited by Aaron Gross and Anne Vallely, 55–78. New York: Columbia University Press.
Newman, Peter. 2014. "Biophilic Urbanism: A Case Study on Singapore." *Australian Planner* 51 (1): 47–65.
Neo, Harvey. 2012. "'They Hate Pigs, Chinese Farmers . . . Everything!' Beastly Racialization in Multiethnic Malaysia." *Antipode* 44 (3): 950–70.
Neo, Harvey. 2016. "Ethical Consumption, Meaningful Substitution and the Challenges of Vegetarian Advocacy." *The Geographical Journal* 182 (2): 201–12.
Neo, Harvey, and J. Z. Ngiam. 2014. "Contesting Captive Cetaceans: (Il)legal Spaces and the Nature of Dolphins in Urban Singapore." *Social & Cultural Geography* 15 (3): 235–54.
Ngiam, Tong Dow. 2007. "Taking over Private Turf for Public's Good." *Today*, 12. February 2, 2007. Accessed October 15, 2019. http://www.wildsingapore.com/news/20070102/070202-2.htm.
Nilsson, Per Ake. 2002. "Staying on Farms: An Ideological Background." *Annals of Tourism Research* 29 (1): 7–24.
Rademacher, Anne, and K. Sivaramakrishnan, eds. 2013. *Ecologies of Urbanism in India: Metropolitan Civility and Sustainability*. Hong Kong: Hong Kong University Press.
Rademacher, Anne, and K. Sivaramakrishna, eds. 2017. *Places of Nature in Ecologies of Urbanism*. Hong Kong: Hong Kong University Press.

Sangma, Sanatombi K. 2016. "Shape-Shifting or Transformation Myth in Garo Culture." *Dialogue: A Journal Devoted to Literary Appreciation* 12 (1): 77–79.

Savage, Victor R. 2019. "Singapore." In *The Wiley Blackwell Encyclopedia of Urban and Regional Studies*, edited by Anthony M. Orum, 1–7. Hoboken, NJ: Wiley-Blackwell.

Savage, Victor R., and Lin Qi Feng. 2020. "Climate Change Adaptation: The Need for an Indian Ocean Regional Metamorphosis." *Journal of the Indian Ocean Region* 16 (1): 6–26.

Singapore Food Agency (SFA). 2019. *Food Farming*. Accessed August 1, 2019. https://www.sfa.gov.sg/food-farming/singapore-food-supply.

Sivaramakrishnan, K. 2015. "Ethics of Nature in Indian Environmental History." *Modern Asian Studies* 49 (4): 1261–310.

Stengers, Isabelle. 2005. "The Cosmopolitical Proposal." In *Making Things Public: Atmospheres of Democracy*, edited by Bruno Latour and Peter Weibel, 995–1003. Cambridge, MA: MIT Press.

Strand, David. 2018. "Singapore's Green Corridor park as a homegrown import." *International Communication of Chinese Culture*. May 8, 2018. 5: 61–81.

Sur, Malini. 2019. "Dreaming Borders: On Cats and Trauma." *Somatosphere: Science, Medicine and Anthropology*. Accessed October 15, 2019. http://somatosphere.net/2019/dreaming-borders-on-cats-and-trauma.html/.

Tan, Audrey. 2016. "Young Farmers Keen, but There're Doubts." *The Straits Times*. June 26, 2016. Accessed October 15, 2019. https://www.straitstimes.com/singapore/young-farmers-keen-but-therere-doubts.

Tan Wei Xian, Alvin. 2010. "Destination Rural: Agri-tainment in the Kranjhi Countryside." Unpublished honours dissertation, Department of Sociology, National University of Singapore.

Tham, Aaron. 2019. "Envisioning Eden: The Manufactured Ecotourism Environment of Singapore." *Journal of Ecotourism* 18 (1): 1–24.

van Dooren, Thom. 2016. "The Unwelcome Crows: Hospitality in the Anthropocene." *ANGELAKI Journal of the Theoretical Humanities* 21 (2): 193–212.

Vaughn, Victoria. "The 'Attack' of Exotic Flora and Fauna in Singapore." *The Straits Times*. November 6, 2010. Accessed October 15, 2019. https://wildsingaporenews.blogspot.com/2010/11/attack-of-exotic-flora-and-fauna-in.html.

Viveiros de Castro, Eduardo. 1996. "Images of Nature and Society in Amazonian Ethnology." *Annual Review of Anthropology*. 25: 179–200.

Viveiros de Castro, Eduardo. 2004. "Exchanging Perspectives: The Transformation of Objects into Subjects in Amerindian Ontologies." *Common Knowledge* 10 (3): 463–84.

Whatmore, Sarah. 2002. *Hybrid Geographies: Natures, Cultures, Spaces*. London: Sage.

Willerslev, Rane. 2007. *Soul Hunters: Hunting, Animism and Personhood among the Siberian Yukaghirs*. Berkeley: University of California Press.

Yee, A. T. K., Richard T. Corlett, S. C. Liew, and Hugh T. W. Tan. 2011. "The Vegetation of Singapore—An Updated Map." *Gardens' Bulletin Singapore* 63 (1 & 2): 205–12.

Yeo, Christa. 2010. "Fresh from the Farms in Singapore." *The Straits Times*. February 21, 2010. Accessed October 15, 2019. https://wildsingaporenews.blogspot.com/2010/02/fresh-from-farms-in-singapore.html.

Yuen, Belinda. 1996. "Creating the Garden City: The Singapore Experience." *Urban Studies* 33 (6): 955–70.

5
The Absent Presence: Potholes in Urban India

Harris Solomon

Introduction: Down the Hole

Urban ecology is a science of permeable dynamics. How might we understand the seasonal human experience of permeable natures in the city? What is present in one moment—the continuity of a road—might be, a month later, washed away by the rain. The politics of material presence can be tricky to track. This chapter presents two arguments stemming from this observation: first, that the biopolitics of cities can be understood as rooted in the seasonal displacements of material absence and presence; and second, that these displacements are visceral and take shape through bodily injury.

It was late August 2017, the peak of monsoon season, and the rains were unusually heavy in Mumbai. The streets flooded and the monsoon eroded a city under constant regimes of land and water system redevelopment. The ecologies of urbanism became headlines (Rademacher and Sivaramakrishnan 2013). Televisions broadcast waters rising in homes and messages to stay safe, and people at work saw these images and tried to go home. It was the unseen that made trouble. A doctor tried driving home from his hospital job. The car stalled, so he left it and attempted to walk through the flooded streets knee-deep with rain. And then, he was gone. He fell into an open manhole, a hole he presumably could not see.

The doctor's (dead) body appeared several days later, several kilometers away. Media reports and everyday discussions centered on the manhole: Why was it open? Who left it open? Accusations flew, some blaming city officials for neglecting to cover these essential portals that drain danger in the island city, but this time one drained a person and a family. The municipality conducted an inquiry that cited eyewitnesses around the area of the manhole. They claimed someone pried it open with a bamboo stick to direct draining, presumably to prevent flood waters entering this person's own home. But this action of preventing water damage to one home damaged the order of life in another. The official inquiry absolved the city government of any wrongdoing (Municipal Corporation of Greater Mumbai 2018). The inquiry report proposed installing nets inside manholes to let water through but to catch bodies should they fall in. The solution to the problem of unruly holes would be a net, a set of orderly holes.

Several days after this death in a manhole, a Mumbai police constable rode home on his motorcycle in the early morning. His bike hit a pothole, a hole he reportedly could not see. His body hit the ground, and his head, too, causing traumatic injuries to his brain. He died in a hospital soon after. The police filed a case of death due to negligence against "unidentified persons responsible for building and maintenance of the road" (Navalkar 2017). Who that is, precisely, remains to be seen (just as many of the holes do). It can be difficult to pin down blame when people move *through* infrastructure, a different sense than "people *as* infrastructure" (Simone 2004, 407).

What is clear is that people call attention to infrastructure's embodied features and organize public life around injury. Holes in the fabric of the city nest in a broader ecology of risk and danger specific to urban environments in South Asia, such that potholes keep a city in a death grip, in ways related to outbreaks of waterborne diseases, habitat-destroying seasonal flooding, and the cyclical disruption and damage of essential services. In this light, what might an embodied urban ecology look like when injury is taken as its central feature? I address these questions based on ongoing ethnographic research project about body-city relations wrought through traumatic injury (Solomon 2017; Solomon Forthcoming). Part of this research entails ethnographic fieldwork at Mumbai's largest public, municipal hospital trauma ward. Inside the hospital I see repeated cases of people encountering holes, especially potholes.[1] Prompted by these cases, in this chapter I direct my inquiry mostly out of the hospital, toward everyday forms of encounter with and discourse around holes in the city. The article describes different forms of hole encounters, including cases of media advocacy, clinical reflections, and attempts by one accident survivor to document danger on the roads. In the conclusion, I consider how the tension between the presence and absence of holes offers broader insight into the politics of public neglect. As it develops an ethnography of holes through these methods, the chapter demonstrates how holes produce a tangled relationship between the trappings of the city and the cavities of living. Being in the

---

1. This chapter stems from a broader research project about the embodiment of infrastructure. Since 2014, I have conducted ethnographic research inside Mumbai's largest public hospital trauma ward, documenting and following cases of traumatic injury from traffic accidents. Many of these cases are due to varying forms of holes, especially potholes. Taking these cases as sentinels of an ever-more normalized material and discursive strain on urban infrastructure, I began to follow cases out of the hospital, and looked for instances where holes produced controversy over injury and death. Inside the hospital I conducted participant observation during different hospital shifts (morning, afternoon, and overnight) to understand different rhythms of injury. Individual interviews with patients or caregivers were tape-recorded when possible, transcribed by me and by a research assistant, and analyzed for emergent concepts and connective themes as the corpus of data grew. Broader context about urban injury came from analyzing city newspaper coverage of health care, transit and traffic politics, and reporting on specific accidents. This was done using database software set to search Marathi, Hindi, and English news sources.

thrall of the urban is one thing, but being in the grip of holes is another, and this grip can kill.

City of Hollows

Holes hollow out the city. On the way to the hospital, my taxi trembles along the pocked surface of a street under construction for the newest phase of the Mumbai Metro, a nearly all-underground passage running north–south in the island city. Thirty meters beneath us, machines burrow tunnels that the Metro will eventually snake through. The Mumbai Metro Rail Corporation, the organization responsible for constructing the Metro, has named each boring machine after India's rivers. The machines "represent force and might, just like a river does," an official noted (Venkatraman 2018). On the side of the street, men stand chest-deep in a hole, also working with force and might. Morning rush-hour traffic snarls around the hole. The men are digging, excavating something. "One government hides gold under the streets, and then the one that follows it digs it up," the taxi driver observes. Whatever the reasons for the holes, the driver says, they continually appear, disappear, and reappear. Contractors get cash and politicians get kickbacks. Signs on the road speak in the present continuous tense and the passive voice: "Work in Progress," or "Inconvenience is Regretted," the intransitive apologia that implicate no one. But there are promises, too, amidst the warnings and apologies. A sign for the next phase of the Mumbai Metro project suggests in large bold script that disruption will lead to something better in the future: "Mumbai is Upgrading!" These are the goings-on of the phrase "work is going on" (*kaam chalu hai*), the everyday expression to describe construction.

Two things are clear from these developments. The first is that holes conjoin politics and affects through structures of inequality (Anand 2017; Bjorkman 2015; Coleman 2017; Cross 2017; Melly 2017; A. Roy 2009; Street 2012). Second, holes in infrastructure are also problems of the body, because they are borne out through injury (De Leon 2015; Jusionyte 2018a, 2018b). If one approach to urban ecologies is through their socio-materiality, my approach here foregrounds how bodies assert ecological specificity. As such, holes share this feature with deadly environments that blur categories of natural and artificial, like borders, but differ in that it is often their absence that is the cause of injury. If one challenge with reckoning with infrastructure is the problem of overcoming language of environmental determinism, as Nikhil Anand notes (2017, 172), with the case of holes, the challenge is reckoning with their unique material form, because they are present-absences. I focus on bodies to do so, and also to reflect on what kind of politics do and do not emerge when wounded bodies gain attention.

The risks that holes pose occur in a context of Mumbai's famous and infamous gridlock traffic, ever intensifying with 250 new cars appearing on the roads each day (Mumbai Metropolitan Region Development Authority 2008). Heavy traffic means that intense vehicle-to-vehicle collisions are low because of slower speeds, but the number of vehicles colliding with pedestrians and motorcycles is very high. Consequently, injuries are the primary cause of death for the 15–24 age group among men in India (Mock et al. 1998; N. Roy et al. 2010). Beneath many of the crashes, lying under the falls and skids and sideswipes, are spaces where the road suddenly changes its grade or composition. Biophysical materialities emerge through their differences: Asphalt turns into gravel; paver blocks end their pattern, leaving the road gap-toothed; jagged edges of potholes fill with water in monsoon season, making it impossible to discern their depth. Wheels of motorcycles hit the holes, filling them in momentarily with rubber. Newspapers, social media, and everyday conversation refer to holes in the road and especially potholes (*gaddhe*) as "death traps." Twitter accounts such as "Ministry of Potholes" serve as a hub for Mumbaikars to showcase crumbles in the roads. Complaints fly at the city's central body of governance, called the *Brihanmumbai Mahanagar Palika* or, in everyday parlance, "the BMC" for its earlier name as the *Bombay Municipal Corporation*.

Later on, laws rush in. It is not quantifiable (yet) amidst the flow of traffic accident data precisely the degree to which potholes or uneven pavement contribute to injury. But that seems beside the point to the most visible advocacy groups in India working on preventing the accident in the first place. This is because public health advocacy groups tend to focus on individual behavior change, with messages to wear a helmet, or wear a seat belt, or drive more slowly. These groups also may pursue national-level legal change and advocacy with police, such as the effort to allow bystanders of the accident ("Good Samaritans") to help the injured with impunity, rather than wrapping them into legal webs that deter their intervention. The road and its holes tend to be secondary on this front. This raises questions about what might be enabling infrastructures to "attack" (Chu 2014, 351) or how they might be understood as "rogue" (Kim 2016, 163). In these cited ethnographic examples—drawn from neighborhood reforms in China and land mines in the South Korean demilitarized zone—the anthropologists insist on the liveliness and generativity of lapses, what Chu terms "the working effects of disrepair" (2014, 353). Holes can be generative, in this light.

Such generative effects take several different forms in India. Amidst the effects of lapses in infrastructure, there is advocacy directed at changing the transit systems of the city. Much of this advocacy work is based on matters of past and potential injury and death, consonant with a biopolitics of damaged livelihood rights (Feldman 2017; Fortun 2009; Redfield 2013; Sunder Rajan 2017).[2] For example, activists and nongovernmental organizations may file a Public Interest Litigation (PIL) that details past injuries associated with a road undergoing long-time unfinished work. Upon inspection of roads in Mumbai that had supposedly been repaired, one report

---

2. In this regard, discussions about deaths in the city from infrastructural holes might be usefully put in conversation with Redfield's notion of "minimal biopolitics" (2013, 18) and Stevenson's discussion of care at a distance (2014, 87).

notes that "engineers were shocked to find entire layers of gravel and stones missing with just a thin film of cement laid on top to hide the shoddy work. A bit like cheap lipstick on chapped, dry lips—it only hides the damage underneath, and that too for a short while" (Despande 2017).[3] The most recent result of scam investigations has been to punish engineers (Singh 2018). The roads stay pocked. But the Bombay High Court's own motions, written by the Court's justices, can be scathing. "Potholes had not been solved from 2016 to 2017," the Chief Justice wrote in a judgment aimed at city authorities (Indian Express 2017). His judgment continued: "How many more people do you want to eliminate till the next monsoon?" (Indian Express 2017). In the eyes of the Court, potholes may not have been eliminated but lives have.

Several insights emerge from these developments. The first is that the materiality of holes is made, not given. The second is that holes are the absence of matter in the wrong place, and that this can be a collective public referent. The third is that holes are somatic. Even as Mumbaikars complain about holes, their bodies appear in them. Holes are deadly, yet they are a rallying cry. Social cohesion rests on sacrifice. At stake is a complex and often dialectical relationship between humans and voids, a relationship that produces hierarchies of holes that ground urban ecosystems. For ecologists, cities *rely* on holes, with permeability as one of their central features. For urbanists, holes define cities: connections of phone lines, drains, and roads are the things that make connectivity possible, to turn isolated forms into the makings of a city system. Connectivity and permeability are essential. The trouble emerges when specific hole-forms emerge as deadly instead of vitalizing.

As law becomes an increasingly common way of reckoning with holes, demands for accountability have formed around holes throughout accounts of governance. For example, the Indian Ministry of Road Transport and Highways counts potholes in its annual report on road accident deaths. The reports use the term "pothole" as one form of classification for uneven surfaces. There are also tabulations for deaths due to other forms of disrepair, such as "loose" surfaces or "road under repair/construction" or "corrugated/wavy road." In sum, these figures add up to the total number of persons killed in road accidents, which are 139,671 deaths in 2014, 146,133 deaths in 2015, and 150,785 deaths in 2016, to take three of the most recently available years since the ministry began publishing its reports in 2008 (Ministry of Road Transport and Highways 2017). With this overflow of accident numbers, whatever the quality of data may be, it is clear that the state sees accounting as a key response to death.[4] Further, it also is clear that accounting can be a site to explore speed and slowdown, too, given how long it may take for one death to register in reports. Mark Lamont explains this point through his research on road accidents in Kenya: "The slowing down of injury compensation cases, the stoppage of traffic at police roadblocks, and the forensic reporting of the media following

---

3. This particular feature of corruption in Mumbai, called "the road scam," has a complex history that is beyond my scope of analysis here because it has taken shape as a set of multiple district court cases.
4. See Nelson (2015) for more on the ways that counting inflects living and dying.

horrific high-casualty crashes all demonstrate the differing paces that regulate roads, drivers and automobiles under the auspices of the centralized state," he notes (Lamont 2013, 382). Holes can stop speed in its tracks and can pause assumptions that acceleration is the only way that injury proceeds.

Killer Potholes

It is December 2017, several months after the doctor disappeared down the manhole, and after the policeman died after his bike hit a pothole. In those months, I have seen numerous cases of head trauma in the hospital trauma ward, often attributed by onlookers or patient relatives to the interference of potholes. I speak with one patient named Asif, who is in the Intensive Care Unit (ICU) following a motorcycle accident. Asif is in his early twenties and studies at a local college. He was on his motorcycle with a friend at night and hit uneven pavement—the edges of a hole. He is among the luckier ones in the ward, as an initial CT scan shows no evidence of traumatic brain injury. His wounds are primarily superficial: Asif bleeds freely from his nose, which is packed with gauze. Compared to the rest of the thirteen patients in the trauma ICU, he's doing quite well. Others with similar injuries will die in the days before and after his time in the ward. Asif will be shifted to a regular ward soon. He is out of danger. The others in beds around him are out of the holes of the street as well. However, for them, danger remains in the form of brain hemorrhages, blunt abdominal trauma, and orthopedic injury. Some will be disabled, and once home, will be thrust into contexts in which disability and identity carry intense semantic and experiential weight (Das and Addlakha 2007; Friedner 2017; Staples 2011).

Asif's friend Nasir stands by his bed. Nasir blames the potholes for the accident. The road was "uneven" and had holes, he said. But what can they do? How to complain against a hole? These roads in Mumbai are "not proper," he said. "I've read the Indian Constitution," Nasir said, and "good roads are a fundamental human right," he thought. But roads-as-rights has yet to appear in the courts, he said, and so there was a hole in the law. If Asif claimed bad road infrastructure as the cause of his accident, no one would listen, Nasir said. The arbiters of things, the police, "only want to hear that Asif was speeding." The onus is on the person on the bike to prove innocence in the face of accusations of speeding or reckless driving. Speed becomes the metric of innocence or guilt. The thing you speed on (or not)—that is, the road—goes scot-free. This demonstrates a silent consensus about the legally inert character of larger vehicles, a point raised by both Jain (2006) in the United States and Lamont (2012, 2013) in Kenya. Yet the type of vehicle matters in this case. All vehicles on the road presumably pass over holes, but certain ones are more likely to get stuck, or send the rider tumbling out, and it is in this moment—the collision—where the fusion of driver guilt and vehicle guilt can occur. There is a seasonality to this, of course: monsoon season erodes pavement and gravel, thinning out the city's skins, abrading the skin of humans. Rhythms get set up around seasons: the holes appear most grievously in monsoon, the municipal government pledges to fix them for the following monsoon, and the cycle repeats.

Stuckedness itself takes shape unequally. Precisely *whose* bodies are exposed to the violence of mobility is thoroughly a matter of gender and sexuality (see Jain 2006). Confinement and movement cleave along lines of gender, class, and kinship, a point elaborated in detail by Phadke, Khan, and Ranade (2011) and Amrute (2015). As Amrute notes, "As women of lower caste and class backgrounds find roads to upward class mobility through pink-collar jobs, they simultaneously are increasingly sexualized, not because they are forced to enter the 'sexually charged' space of the car and the street, but because these very spaces increasingly figure as the borderlands of a respectable middle-class imaginary of consumption taking place behind office park and residential compound walls" (2015, 341). The holes of the street constrict and open according to multiple parameters of social structure. The fact that Asif and Nasir are both young men is hardly incidental, as it is the case that most of the motorcycle accidents in the trauma ward occur among young men. It is also notable that when women are in the trauma ward due to motorcycle accidents, they are usually cases of pillion riders. (Mumbai's laws do not require pillion riders to wear helmets.) Exposure to mobility and its entailments can mean opportunity and it can mean vulnerability; both forms are always in relation to different concentrations of social value.

The same week as Asif's admission to the trauma ward, there is news of a motorcyclist who has died because of a pothole. The motorbike driver was a doctor who coached chess for kids in his free time. He was traveling to pick up some trophies for a competition. He died on the freeway. He attempted to avoid a pothole, swerved out of his path, and was hit by a truck moments after. A Marathi newspaper describes the street as rough and uneven (*khadbadit*). The pillion rider described the bits he remembered of the incident. The doctor "spotted a large pothole in our way at the last minute and slightly lost control over the vehicle. Just then, the *tempo* [truck] hit us from behind. I don't remember what happened after that." A news anchor covering the story rued that "another life is lost to killer potholes" (*Times of India* 2017). (No word on the truck.) The anchor called the accident a case of "civic negligence," raising again the common concern of many in this city, even amidst media hype: How is it that the city is a form of livelihood yet also a form of death? Here, "the civic" is not so much a duty but a form of deadly neglect: the commons that kills, and does so through gaps.[5] The locations of the gaps tell their own story about how injuries embed in a context of several transitions of capital flow during India's liberalization, neighborhood-level social class changes, and tectonic shifts in real estate markets (Appadurai 2000; Finkelstein 2015; Rao, Pemmaraju, and Pietrusko 2007; Searle 2016).

---

5. On the ambivalence of the commons, see Berlant (2016).

How might one understand the ways that holes materialize both neglect and profit? Anthropologist Filip De Boeck elaborates this idea at length based on research in Kinshasa, where the material-theoretical concept of a *hole* organizes social formation and deformation. Potholes are a key example. "Postcolonial urban living in Congo literally means living with potholes as generic urban infrastructures," he writes (2016, 13).[6] Potholes are a matter of possibility, something generative. I agree with De Boeck on the generativity of potholes and that they are scenes of instruction. But we can also see how people confront the ways that potholes generate death and injury, and how publics emerge through life's erosion. Although people in Mumbai don't necessarily cast the city as a giant hole, the way that De Boeck suggests that residents of Kinshasa do, the hole is a more a semio-material figure of neglect. As the following section describes, whether they are rendered comedic or anchored in the putatively flat reason of a "citizen audit," holes point to a kind of negligence that perhaps someone will respond to.

City Songs

The video is a music video, although not one of any established band. It is a group of people who work at Red FM, one of the city's most popular radio stations. The camera centers on a woman well known to Mumbaikars, the radio jockey (RJ) named Malishka Mendonsa. Mendonsa begins singing, with the group following behind her in chorus. She addresses the city itself, and its relationship to potholes. The song works through the register of trust (*bharosa*)—or, more precisely, lack thereof—in the city's governing body: the BMC. The first line, and the structure of the refrain, affirms that the city doesn't trust the municipal authorities. This is especially pronounced during monsoon season when infrastructure collapses. It is an urban anthem, one that voices the common man's mistrust of governance, pitted like the potholes in the roads, round and full as the pothole's own shape. It was posted in July 2017; in the six months that followed, it had over 22 million views on YouTube. The song, sung in Marathi, is a play on a Marathi folk song that works similarly through rhyme and reduplication of words (the original song is called *Sonu, Tuzha Mazhavar Bharosa Nahin Kay* [Sonu, Don't You Trust Me]).

In the song, Mendonsa is the siren, with the city speaking to and through her. The first two stanzas sketch out the antics (*jhol*) around potholes:

> *Mumbai, tula BMC var bharosa nahin kay*
> Mumbai, don't you trust the BMC
> *Mumbai cha rasta madhye jhol jhol*
> There are some antics on Mumbai's roads
> *Rastyanche khade kase khol khol*

---

6. De Boeck contrasts his use of "hollow" here with that of Weizman (2012); the hollowness of Kinshasa "should be understood in a much more immediate" way, De Boeck notes (2016, 16).

The potholes in the road are deep
Khadyancha aakar kasa gol gol
The potholes are round-shaped
Mumbai tu maya sang goaḍ bol
Mumbai, you speak sweetly to me
Mumbai tula pausa war bharosa nahin kay
Mumbai, don't you trust the rains
Mumbai cha traffic kiti laamb laamb
Mumbai's traffic is so long
Traffic madhye aapan jaam jaam
We're stuck in traffic jams
Signal cha aakar kasa gol gol
The traffic signal is round-shaped
Mumbai tu maya sang goaḍ bol
Mumbai, you speak sweetly to me.

(*Red FM India* 2017)

The song continues with images of overflowing rains in monsoon season and of delayed local trains. It concludes on two parallel notes. One is of exhaustion, of the common man (*manus*) drained by the potholes and the breakdown. The body count of the rhythms of urban ecology become too much to bear, leading to a sense of an overwhelmingness of the city that is not borne out of the shock of crowds but, instead, out of the shock of the seasonal yet seemingly endless effects of infrastructural attack. The other conclusion is that these holes are evidence. The rains wash away the patina of shoddy patchwork on potholes, and potholes materialize municipal corruption. The song marked another moment that followed several acts of pothole publicity via Mendonsa's radio show. In 2013, she hosted a "Pothole Festival" (*pothole utsav*), inviting listeners to photograph, describe, and report potholes on Mumbai's roads on the radio station's Facebook page. In turn, Mendonsa went to some of these potholes to do a pothole *puja*, or act of worship, with offerings of flowers and songs to the pothole. The radio station invited local municipal government officials to join, in part to fold them into a field of accountability.[7] In the pothole *pujas*, public acts of devotion cluster around the failings of a city. In the case of potholes, however, parody is as important a structure and aesthetic as is Hindu devotion. Everyone is in on the joke, including the government officials.

The hole also anchors a transition space for the meeting of two parties—citizens and their municipal representatives—who are otherwise often separated by endless phone calls and written complaints that may or may not get answered. As Jennifer Ashley notes in her ethnography of parody on Chilean television, the transitional effects of parody can form "sites for the emergence of new political actors and practices" (2014, 767). Yet, she notes, ethnographic focus on what precedes or follows a parodic transition "can allow us to skip too quickly over what is created in

---

7. Pothole *pujas* also occurred in Bangalore.

Figure 5.1: Pothole "puja" (worship)

the in-between" (2014, 767; see also Boyer 2013; Haugerud 2013). If neglect creates holes, and holes create damage, what and who lie in the middle?

I suggested earlier that bodies lie in the holes and are pulled out for treatment in hospitals. But there are deliberate insertions of bodies into potholes too. One in Bangalore, for example, drew attention to the ever-present feature of potholes in the city through performance art. Dressed as a mermaid, she sat in a water-filled pothole and splashed herself as the cameras clicked. Potholes become places to reimagine presence for others as well. Architect Rupali Gupte created a project titled "Pothole City," in which she constructed a miniature town inside a pothole. For Gupte, the pothole is a sign of urban inhabitation, a sign that people are living in a place, treading through it. The hole is not necessarily deleterious. It can be a rebuke to authoritative master plans, a future that Gupte notes emerges "from the madness of the city and not from the logic of the grid" (2011).

Technology firms have also found potential in potholes. An app called Spothole by a Mumbai-based software developer group offers users the ability to take photos and geo-locate potholes. In turn, other app users can take this information into account as they navigate the same streets. It's holesourcing, where the crowd that might participate and know the digital urban commons is anchored in moving around the holes instead of into them. The app designers wanted to puncture complicity among citizens, fed up with municipal governance fractures but also too used to it to do much. "All roads lead to potholes. The authorities turn a blind eye to the situation on the ground. We, the people of Mumbai, eventually learn to live with it and move on" they announce (Fill in the Potholes Project 2014). The ads for the app feature miniature toys inside potholes. Marking out potholes, the app developers suggest, is one way out.

Figure 5.2: Image of artist Rupali Gupte's installation "Pothole City"

Acts of publicity around potholes share features with acts of "nuisance talk" (Ghertner 2015, 80) and imperatives to clean the city in order to expand middle-class access while rooting out the poor (Anjaria 2009; 2016). And although the *pujas* are centred on potholes, one cannot overlook the act of ritual itself, one of the critical features of divine spaces in urban South Asia (Benjamin 2015; Srinivas 2006; Taneja 2017). The emphasis on potholes in the songs, the *pujas*, the artistic forms, and the performativity of tongue-in-cheek parody each reveal different valences of the pothole. The hole can be a megaphone, a channel for urban outcry. It can hold art, and imagination, and ritual, in the forms of mermaids or miniatures. These are valences where the hole holds life. Yet holes have an underside, the bottom surface of the hollow space. The different pothole groups offer a common refrain: they do their work because potholes kill. You can be trapped in them, and so can your life. The hole may be a possibility for remedy, but it has an original violence that lingers, a remainder that is not easily sung away.

Grievance Portal

Bipin and his motorbike hit the pothole at the same time. Luckily, he stayed on the bike. Not everyone is so lucky. He stopped and lay down on the road in what he described as "excruciating" pain—"it left a chill in my spine," he said to me. His incident happened on his daily college commute, on a major arterial road in the Mumbai suburbs. It's a normal sort of story: a commute, a pothole, and an injury that seemed minor. With lingering back pain, Bipin returned to his pothole scene a few days later and photographed the pothole. And then he began photographing others, taking photos on his mobile phone. He went up and down the length of the arterial road for several kilometres, geocoded each pothole photo with the help of Google Maps, and compiled a report. He is studying journalism in college, and felt this was a chance to put his reporting skills to the test. Bipin filed his report on the city's online site for municipal complaints (*Aaple Sarkaar*, "Your Government"), an e-governance "grievance portal" for the BMC.

Bipin's report begins with a letter, a plea: "Hoping for a response soon with actions taken to improve the condition of the roads as soon as possible before someone is left severely injured." Each page conveys a pothole picture and sometimes also a vignette because Bipin interviewed nearby figures like the chai seller or the newspaper vendor. They recalled accidents at the given pothole, a string of memories pegged to the pavement's unevenness. This was the grounds of the evidence of urban neglect. One vendor told Bipin that he took it upon himself to fill the holes in front of his newsstand with extra sand he had from his own home repair work. But the holes kept reappearing. He told Bipin that a pit was created every fifteen or twenty days. The news vendor regularly points out the holes to the traffic police, he said, but they affirm that holes are the municipality's responsibility, not theirs. The accusation of negligence takes shape, and then bounces.

But sometimes the complaint can bounce back through the hole. Over a week after filing the report, Bipin received a response that his complaint had been resolved. He returned to the potholes to check. The one he had mentioned in the report where his bike skidded had been patched over, but unevenly so. The others had barely been touched, or if there was a touch, it was a touch-up, a patch of gravel or asphalt patted down. "People can fall in those," he told me. A patch is not a fix, and the patch can trip you up. "See, the base isn't proper," he explained. "They just filled it up"—and did so often quite partially. Filling in isn't the same as fixing. The quick deadlines the city sets for contractors are part of the issue, he thought. For him, the very return to the scene of injury was traumatic. He had a mental picture of "the worst" when he got back to that original hole, a picture in his mind of him lying on the road crouched in pain but maybe not ever to get up again. He went to audit, a supposedly distanced act between him and the pothole. But he realized he and his body still had a connection to it.

The failure of an appropriate civic response led Bipin to the news media, an outlet many urban Indian (and often middle-class) citizens pursue. A news story ran his account of injury, photographic capture, pothole mapping, and the feeble civic response. After a decent run of Facebook posts and retweets and shares of the news story on WhatsApp, Bipin went to the headquarters of the municipality. The Chief of the BMC was not present. Bipin left his report with the secretary. The "refusal of the municipality to listen is costing people their lives," he said. The grounds of negligence here is not listening: the trap of silence.

The accident reinforced his own expectation of neglect in the form of the holes that the city produces. This affects how he rides now on his motorcycle. "I have a mental fear of negligence" by the city authorities, Bipin claimed. "You're so sure that the BMC hasn't filled in the potholes, so you slow down [your motorcycle], but vehicles may hit you from the back." He feels negligence, and it changes the way he drives his motorcycle, even if that change can cause a multivehicle pile-up. For Bipin, negligence is something that leaves a lasting effect, something that works phenomenologically.

I took this bodily dimension of negligence—that Bipin still *feels* the accident—as essential to his call for a response from civic authorities. And yet, that response was uneven, pointing out perhaps the limits of politics of wounds and wounding that assert urban ecological materialities. There was no lack of bodies in Bipin's case: The photos of the potholes were photos of holes that *could* contain a body, should misfortune and bad timing line up. Or, as he had put it, Bipin was well aware that filing a report online might have little effect. He was aware of his position as a "compliant consumer-citizen," to use a term for the subject position offered by William Mazzarella in an essay about the politics of e-governance in India (2006). Mazzarella notes that in India, long rued for its corruption, the possibilities offered by e-governance platforms defy easy categorizations of care or neglect. "The provision of a channel of expression, irrespective of actual government response or action, was in itself a therapeutic gesture," Mazzarella writes (2006, 486). And it was in here, somewhere amidst such a promise, that Bipin was stuck (although he managed to survive it). Bipin untangled himself from the hole in the road, only to find himself stuck in the grievance portal. Perhaps this is characteristic of his lower-middle-class background, given the ways that media outcry both produces and is produced by middle-class status. What was most important to him was less his own class position and more his belief that it would likely take the death of someone far more socially high-profile before anything significant might be done by the authorities. "They wait for a major incident to occur, only if a celebrity or politician dies," he said. Then the hole might get filled, properly. Until then, it was a cycle of fill, dissolution, complaint, and refill, on permanent repeat.[8] Bodies may be the centerpiece of asserting biophysical dangers, but that does not necessarily mean they are the most convincing forms of political tactics to gain redress.

Bipin took particular pride in his report not only for its close attention to geocoded detail but for the photographs themselves. The science of it all, the precision, came from the frames he shot that placed the holes in context. A Google Maps coordinate was certainly specific, he thought, but knowing that a pothole appeared in front of a specific tea stall or storefront made it just as real. His photos are multiscaled in the report: Some are close-ups of the edges of holes, at their level, a hole's-eye view of things. Others are taken from the perspective of a motorcyclist, with the camera angled several feet above the street at a downward angle. I mentioned that he had a knack for photography.

---

8. In February 2018, the Bombay High Court asked the BMC to develop a "portal" online specifically for reporting potholes.

He brought out his phone to show me his favorite photograph so far. He had photographed a bus stuck in the water, in the heavy rainfall of September 2017, the same day that the doctor fell down the manhole. Bipin saw a bus get stuck inside an underpass tunnel, as the waters rose. He went closer, even though people nearby warned him to stay back. But he was concerned that people were stuck in the bus, so he waded through the water. Monsoon has heroic moments like this, moments when people make efforts at rescue that later get called "Mumbai's spirit" in a refrain of resilience, another story that the city tells itself.

"Were you alone?" I asked. No, he said, a friend was with him. But the friend would only go so far, and after a point, Bipin was alone in his attempt to free anyone stuck in the bus.

"But why were you alone?" I asked. "Why did the friend not join you?"

"He was afraid of getting sucked into a hole."

Conclusion: The Hole of Neglect

As injury and death accumulate on Mumbai's roads, this chapter has explored holes as fleshy, experiential, and discursive sites through which infrastructural failure generates and deforms sociality through bodies. In Mumbai's holes, there is much that proliferates even as life degrades, and seasonal rhythms to this process have the at-times contradictory effect of making it seem like the damage won't stop, even as people know that at some point the rains will cease. That tension has been central to the encounters described here and highlights how bodies are inextricable from infrastructure and urban ecological biopolitics more generally. I have suggested that holes are things of somatic encounter, and the somatic encounter with infrastructure's absence points to lapses in law, governance, and civic attention. These complexities are socio-spatial and distribute unevenly. Bodies in holes exemplify the complexities of injury and recovery that make infrastructure biopolitical, and bodily injury is telling about a city's multiple ecologies, such that life and death themselves should be understood as seasonal forms. Akhil Gupta (2012) writes of "uncaring" as a political mode of bureaucracy that illustrates how the power of the Indian state can take shape as presence through absence. In the cases I have described, politics is provision with holes, a kind of daily mobility with punctures. Holes reshape care because they materialize *both* presence and absence. This tension between presence and absence is a biopolitical form that merits deeper ethnographic attention. Holes demand action, but the demand is often only addressed in death. I largely watch this unfold from the inside of the trauma ward, where the cases arrive after buildings collapse on people, or after bridge stampedes and suffocation deaths from fires attest to the absence of holes. These events resulting from the absence of holes fill Mumbai's casualty wards, which are already filled with bodies injured by holes that are excessively present.

I am in a taxi again. As usual, the car bumps along the road. The radio jockey Malishka chatters on the car's radio. It is not monsoon season yet, but meteorologists have predicted aberrant rain showers. "Are we prepared for the rains?" Malishka asks listeners—in this case, the driver and me. She continues: "And are we ready for potholes? Ninety-foot-deep potholes?" I ask the driver about potholes. They're like the clouds, he says, pointing to the gray sky. They come and go, these clouds of the road, feathery from afar, jagged up close.

Works Cited

Amrute, Sareeta. 2015. "Moving Rape: Trafficking in the Violence of Postliberalization." *Public Culture* 27 (2): 331–59.

Anand, Nikhil. 2017. *Hydraulic City: Water and the Infrastructures of Citizenship in Mumbai.* Durham, NC: Duke University Press.

Anjaria, Jonathan Shapiro. 2009. "Guardians of the Bourgeois City: Citizenship, Public Space, and Middle-Class Activism in Mumbai." *City and Community* 8: 391–406.

Anjaria, Jonathan Shapiro. 2016. *The Slow Boil: Street Food, Rights, and Public Space in Mumbai.* Stanford, CA: Stanford University Press.

Appadurai, Arjun. 2000. "Spectral Housing and Urban Cleansing: Notes on Millennial Mumbai." *Public Culture* 27: 627–51.

Ashley, Jennifer. 2014. "Prime-Time Parody: News, Parody, and Fictional Credibility in Chile." *American Ethnologist* 41 (4): 757–70.

Benjamin, Solomon. 2015. "Cities within and beyond the Plan." In *Cities in South Asia*, edited by Crispin Bates and Minoru Mio, 98–122. New York: Routledge.

Berlant, Lauren. 2016. "The Commons: Infrastructure for Troubling Times." *Environment and Planning D: Society and Space* 34 (3): 393–419.

Bjorkman, Lisa. 2015. *Pipe Politics, Contested Waters: Embedded Infrastructures of Millennial Mumbai.* Durham, NC: Duke University Press.

Boyer, Dominic. 2013. "Simply the Best: Parody and Political Sincerity in Iceland." *American Ethnologist* 40 (2): 276–87.

Chu, Julie. 2014. "When Infrastructures Attack: The Workings of Disrepair in China." *American Ethnologist* 41 (2): 351–67.

Coleman, Leo. 2017. *A Moral Technology: Electrification as Political Ritual in New Delhi.* Ithaca, NY: Cornell University Press.

Cross, James. 2017. "Off the Grid: Infrastructure and Energy beyond the Mains." In *Infrastructures and Social Complexity: A Companion*, edited by Penelope Harvey, Casper Jensen, and Atsuro Morita, 198–210. New York: Routledge.

Das, Veena, and Renu Addlakha. 2007. "Disability and Domestic Citizenship: Voice, Gender, and the Making of the Subject." In *Disability in Local and Global Worlds*, edited by Benedicte Ingtad and Susan Reynolds Whyte, 128–48. Berkeley: University of California Press.

De Boeck, Filip. 2016. *Suturing the City: Living Together in Congo's Urban Worlds.* London: Autograph ABP.

De Leon, Jason. 2015. *The Land of Open Graves: Living and Dying on the Migrant Trail.* Berkeley: University of California Press.

Despande, Tanvi. 2017. "Probe Finds around 100 Engineers Guilty in Road Works Scam." *Mumbai Mirror*. December 28, 2017.
Feldman, Ilana. 2017. "Humanitarian Care and the Ends of Life: The Politics of Aging and Dying in a Palestinian Refugee Camp." *Cultural Anthropology* 32 (1): 42–67.
Fill in the Potholes Project. 2014. "Let's Fill in the Potholes." YouTube. October 4, 2014. Accessed May 5, 2018. https://www.youtube.com/watch?v=mWjVKUGn4Cs.
Finkelstein, Maura. 2015. "Landscapes of Invisibility: Anachronistic Subjects and Allochronous Spaces in Mill Land Mumbai." *City & Society* 27 (3): 250–71.
Fortun, Kim. 2009. *Advocacy after Bhopal: Environmentalism, Disaster, New Global Orders*. Chicago: University of Chicago Press.
Friedner, Michele. 2017. "Sign Language as Virus: Stigma and Relationality in Urban India." *Medical Anthropology* 37 (5): 359–72.
Ghertner, Asher. 2015. *Rule by Aesthetics: World-Class City Making in Delhi*. New York: Oxford University Press.
Gupta, Akhil. 2012. *Red Tape: Bureaucracy, Structural Violence, and Poverty in India*. Durham, NC: Duke University Press.
Gupte, Rupali. 2011. "Pothole City." *A Provisional Practice*. June 28, 2011. Accessed May 5, 2018. http://aprovisionalpractice.blogspot.in/2011/06/pothole-city.html.
Haugerud, Angelique. 2013. *No Billionaire Left Behind: Satirical Activism in America*. Stanford, CA: Stanford University Press.
Indian Express. 2017. "Potholes: Bombay High Court Seeks Response from State Municipal Bodies, Councils." September 29, 2017.
Jain, Lochlann. 2006. *Injury: The Politics of Product Design and Safety Law in the United States*. Princeton, NJ: Princeton University Press.
Jusionyte, Ieva. 2018a. "Called to 'Ankle Alley': Tactical Infrastructure, Migrant Injuries, and Emergency Medical Services on the US-Mexico Border." *American Anthropologist* 120 (1): 89–101.
Jusionyte, Ieva. 2018b. *Threshold: Emergency Responders on the US-Mexico Border*. Berkeley: University of California Press.
Kim, Eleana. 2016. "Toward an Anthropology of Landmines: Rogue Infrastructure and Military Waste in the Korean DMZ." *Cultural Anthropology* 31 (2): 162–87.
Lamont, Mark. 2012. "Accidents Have No Cure: Road Death as Industrial Catastrophe in Eastern Africa." *African Studies* 71 (2): 174–94.
Lamont, Mark. 2013. "Speed Governors: Road Safety and Infrastructural Overload in Postcolonial Kenya, c. 1963–2013." *Africa* 83 (3): 367–84.
Mazzarella, William. 2006. "Internet X-Ray: E-Governance, Transparency, and the Politics of Immediation in India." *Public Culture* 18 (3): 473–505.
Melly, Caroline. 2017. *Bottleneck: Moving, Building, and Belonging in an African City*. Chicago: University of Chicago Press.
Ministry of Road Transport & Highways Government of India (2017). Annual Report 2016–17, New Delhi, India.
Mock, Charles, Gregory Jurkovich, David nii-Amon-Kotei, Carlos Arreola-Risa, and Ronald Maier. 1998. "Trauma Mortality Patterns in Three Nations at Different Economic Levels: Implications for Global Trauma System Development." *The Journal of Trauma and Acute Care Surgery* 44 (5): 804–14.

Mumbai Metropolitan Region Development Authority (2008). Comprehensive Transport Study, Mumbai, India.

Municipal Corporation of Greater Mumbai. 2018. Enquiry report on the incident of demise of Dr. Amarapurkar due to falling in S.W.D. manhole on August 29, 2017.

Navalkar, P. 2017. "Constable Dies a Week after Skidding on Pothole." *Asian Age*, September 8, 2017.

Nelson, Diane. 2015. *Who Counts: The Mathematics of Death and Life after Genocide*. Durham, NC: Duke University Press.

Phadke, Shilpa, Sameera Khan, and Shilpa Ranade. 2011. *Why Loiter? Women and Risk on Mumbai's Streets*. New Delhi: Penguin.

Rademacher, Anne, and K. Sivaramakrishnan. 2013. *Ecologies of Urbanism in India: Metropolitan Civility and Sustainability*. Hong Kong: Hong Kong University Press.

Rao, Vyjayanthi, Satya Pemmaraju, and Gerard Pietrusko. 2007. "Venture Capital." *Public Culture* 19 (3): 593–609.

Red FM India. 2017. "Sonu Song Pothole Mix with Malishka | Mumbai Tula." YouTube. July 10, 2017. Accessed May 5, 2018. https://www.youtube.com/watch?v=U5npFH8v8a4.

Redfield, Peter. 2013. *Life in Crisis: The Ethical Journey of Doctors Without Borders*. Berkeley: University of California Press.

Roy, Ananya. 2009. "Civic Governmentality: The Politics of Inclusion in Beirut and Mumbai." *Antipode* 41 (1): 159–79.

Roy, Nobhojit, V. Murlidhar, Ritam Chowdhury, Sandeep Patil, Priyanka Supe, Poonam Vaishnav, and Arvind Vatkar. 2010. "Where There Are No Emergency Medical Services: Prehospital Care for the Injured in Mumbai, India." *Prehospital and Disaster Medicine* 25 (2): 145–51.

Searle, Llerena. 2016. *Landscapes of Accumulation: Real Estate and the Neoliberal Imagination in Contemporary India*. Chicago: University of Chicago Press.

Simone, AbdouMaliq. 2004. "People as Infrastructure: Intersecting Fragments in Johannesburg." *Public Culture* 16 (3): 407–29.

Singh, D. 2018. "Road Scam: 13 Engineers to Appeal Against Punishment." *Indian Express*, March 7, 2018.

Solomon, H. Forthcoming. *Lifelines: The Traffic of Trauma*. Durham, NC: Duke University Press.

Solomon, H. 2017. "Shifting Gears: Triage and Traffic in Urban India." *Medical Anthropology Quarterly* 31 (3): 349–64.

Srinivas, Tulsi. 2006. "Divine Enterprise: Hindu Priests and Ritual Change in Neighbourhood Hindu Temples in Bangalore." *South Asia* 29 (3): 321–43.

Staples, James. 2011. "At the Intersection of Disability and Masculinity: Exploring Gender and Bodily Difference in India." *Journal of the Royal Anthropological Institute* 17 (3): 545–62.

Stevenson, Lisa. 2014. *Life Beside Itself: Imagining Care in the Canadian Arctic*. Berkeley: University of California Press.

Street, Alice. 2012. "Affective Infrastructures: Hospital Landscapes of Hope and Failure." *Space and Culture* 15 (1): 44–56.

Sunder Rajan, Kaushik. 2017. *Pharmocracy: Value, Politics and Knowledge in Global Biomedicine*. Durham, NC: Duke University Press.

Taneja, Anand Vivek. 2017. *Jinnealogy: Time, Islam, and Ecological Thought in the Medieval Ruins of Delhi*. Stanford, CA: Stanford University Press.
Times of India. 2017. "Pothole Claims Another Life: Mulund Doctor Run Over by Truck." December 9, 2017. Accessed May 5, 2018. https://timesofindia.indiatimes.com/videos/city/mumbai/pothole-claims-another-life-mulund-doctor-run-over-by-truck/videoshow/61995379.cms.
Venkatraman, T. 2018. "Tunnel-Boring Machines Used for Mumbai Metro Work Named after Rivers." *Hindustan Times*, February 1, 2018.
Weizman, Eyal. 2012. *Hollow Land: Israel's Architecture of Occupation*. London: Verso.

6

The Death and Life of Urban Ecological Commons in Taipei

Tomonori Sugimoto

One spring day in 2018, I was with Palafang in her garden. I had met this indigenous woman at Icep, a public housing complex a few kilometers away, located in the eastern part of New Taipei City, Taiwan.[1] We were standing on a tract of hillside land. I looked around our surroundings. The hill toward our south was thickly covered with trees. A small creek was running below us, and toward our north I could see freeway overpasses and high-rises from a nearby science park. Accompanied by her dog, we talked about various vegetables she was growing in her plot—bitter melons, scallions, and yams. I recognized some plants central to her Pangcah/'Amis foodways, like *tayalin* (*Solanum integrifolium*), small and bitter eggplants so named because they are shaped like tires.[2] In the corner of her garden, Palafang had built a *taloan*—a small wooden hut—with recycled wood so that she can take occasional breaks while working there. She told me that she usually comes here for a few days in a row, and after picking up enough vegetables, she sells them on the street near where she lives.

Few Taipei residents would know that this hillside land exists. To get there that day, I hired a taxi from the downtown area near Icep. After driving toward the hilly and less populated southern side of the neighborhood, for several minutes we were on a narrow street adjacent to rail tracks designed only for one-way traffic. After

---

1. Icep is a pseudonym. In this chapter, I reserve the term "indigenous" (as well as "native") for Taiwan's Austronesian peoples who are categorized as, and identify as, *yuanzhumin* (literally meaning "those who are indigenous on the island" in Mandarin). Descendants of Han Chinese people who have settled from mainland China in multiple waves since the seventeenth century also have come to claim localness to the island, especially as what scholars call Taiwanese consciousness saw a rise after the fall of KMT authoritarian rule in 1987. However, these people do not identify as indigenous. They instead ethnically identify as "Han people" (*hanren*). Many also identify as "Taiwanese people" (*taiwanren*), rather than "Chinese people" (*zhongguoren*).
2. *Taya*, in her indigenous Pangcah/'Amis language, is a loanword from Japanese. In English, it is also called "ornamental pumpkin" or "pumpkin on a stick" (erroneously, because they are actually in the eggplant family). In Mandarin Chinese, the Pangcah/'Amis are collectively called *ameizu* (*zu* meaning tribe, while *amei* is a phonetic translation of 'Amis). Those from the Taitung region tend to call themselves 'Amis, while those from Hualien identify as Pangcah. Both the Pangcah from Hualien and the 'Amis from Taitung are represented in most urban communities, hence when I refer to urban community residents collectively, I use "Pangcah/'Amis" to acknowledge this regional diversity. When an informant's geographical origin and identification is clear, I will use either Pangcah or 'Amis.

Figure 6.1: Gardens on a hillside in New Taipei City

driving past old houses, a Taoist temple, and freeway overpasses, the street ended and the hillside spread before my eyes. On my initial visit there with some close friends from Icep a few years before, the hillside seemed like an archetypal urban space neglected by the state. But on my repeated visits there I realized that it was a very valuable piece of land from the perspective of people who utilize it despite their not holding formal property rights. Both sides of the creek running through the land have multiple gardens (Figure 6.1). Their owners, like Palafang, are mostly indigenous people living nearby. Many of them diligently come here every day on their scooters or electric bicycles, in the wee hours before the sun comes out, and return home after tending to their plots for a few hours. Over the course of my ethnographic research in the Taipei metropolitan region since 2014, I have come to know many indigenous people using the city's underutilized public lands like hillsides and riverbanks. It is in such little-noticed gaps of the city—which I call *urban ecological commons* in this chapter—that native people who have settled in this metropolis enact a wide range of relations with more-than-human-beings—catching fish and shrimp, picking wild plants, and growing vegetables important in their indigenous foodways.

In this chapter, I examine the death and life of such urban ecological commons in the Taipei metropolitan region. In conceptualizing urban ecological commons, I am indebted to Vinay Gidwani and Amita Baviskar's (2011) short and provocative essay, "Urban Commons," as well as Anna Tsing's 2015 book, *The Mushroom at the End of the World*. Gidwani and Baviskar conceptualize urban commons as follows:

Urban commons include so-called "public goods": the air we breathe, public parks and spaces, public transportation, public sanitation systems, public schools, public waterways, and so forth. But they also include the less obvious: municipal garbage that provides livelihoods to waste-pickers; wetlands, waterbodies, and riverbeds that sustain fishing communities, washerwomen, and urban cultivators; streets as arteries of movement but also as places where people work, live, love, dream, and voice dissent; and local bazaars that are sites of commerce and cultural invention. (Gidwani and Baviskar 2011, 43)

It is this second, less obvious, and less formal type of urban commons that interest me here. In this chapter, I specifically focus on urban *ecological* commons, paying attention especially to commons whose functions are ecological and opening up the concept of commons to include a variety of nonhuman beings (see also Tsing 2015, 255). In the context of Taipei, I conceptualize public lands near rivers and on hillsides as such urban ecological commons, where marginalized people "ignore questions of property" (Tsing 2015, 78) to work with land, acquire food items like plants, and even alleviate a sense of alienation in the city. In that sense, urban farming and gardening practices that I examine in this chapter differ from what other authors in this volume examine, such as visits to highly commodified agri-tourism farms in Singapore (see Annu Jalais' chapter)[3] or organic farming that middle-class urbanites practice on privately owned rooftops and terraces in Bengaluru, India (see Camille Frazier's chapter).[4] Gardening and farming practices I am describing here are more guerrilla (e.g., Bayat 2013), tied not to privately owned spaces but to urban ecological commons (at least theoretically) open for use and appropriation by anyone.

Paying attention to urban ecological commons also allows me to expand our discussion of commons in anthropology and its related disciplines. Scholars in recent years have focused mostly on urban commons as overt sites of political action, like the Occupy Movement (e.g., Holston 2019; Susser 2017). Urban ecological commons in Taipei have certainly incited such political actions, especially movements against forced relocation, as we will see below. However, urban ecological commons have political potential even beyond what we typically see as politics (like protests). The persistence of urban ecological commons through everyday engagement like gardening and foraging can challenge dominant ideas about land, nature, and the city, albeit in more covert fashion.

Despite their importance to some urban dwellers, urban ecological commons are quickly disappearing in Taipei, due to massive development projects and state-led greening of the city in the last few decades (other contributors to this volume

---

3. See Annu Jalais, "The Singapore 'Garden City': The Death and Life of Nature in an Asian City," chapter 4 in this volume.
4. See Camille Frazier, "Putting the Garden Back: Cultivating Life through Urban Gardening in India," chapter 2 in this volume.

like Caroline Merrifield[5] and Erik Harms[6] have also noted the impact of similar processes in Hangzhou and Saigon, respectively). The city's abundant hillsides have been flattened to make way for middle-class housing complexes and gated communities, which tout the proximity to "nature" as their selling point; riverscapes have been transformed into sites of leisure, equipped with parks and bicycle paths, which were built by displacing both humans and nonhumans. The ecologies of urbanism analytic is helpful here to elucidate how urban natures are "given social meaning in ways that compel human action"; actions that are mobilized in the name of urban natures also include violence, exclusion, and displacement (Rademacher and Sivaramakrishnan 2017, 11). In the context of Taipei, I am interested in how native people's relations to urban ecological commons have been challenged but also linger—highlighting the forced death and persistent life of nature in one Asian metropolis.

Despite violent attempts by the Taiwanese state and capital to impose certain ways of ecological belonging, however, they have failed to eliminate other ways of knowing and experiencing Taipei's lands and natures. Urban ecological commons persist. Different visions of being green in the city are constantly pushing against each other in this city. As Gidwani and Baviskar put it, urban ecological commons "thrive and survive by dancing in and out of the State's gaze, by escaping its notice, because notice invariably brings with it the desire to transform commons into state property or capitalist commodity" (2011, 42). While Anna Tsing does not specifically address *urban* commons, she too notes that commons are "latent" in that they are largely invisible, escaping notice, without formal institutionalization (2015, 255). To trace the death and life of urban ecological commons, I will rely on my ethnographic research with indigenous Pangcah/'Amis people living in the Taipei region since 2014.[7] Like other place-based movements examined by Arturo Escobar, Dianne Rocheleau, and Smitu Kothari, in their cosmology "places are not treated as real estate, as exchangeable and interchangeable commodities, but rather as the ground where body, home, community and habitat are joined in everyday experience as well as in history" (2002, 35). They look at Taipei's riverbanks and hillsides not first and foremost as commodities, waiting to be exchanged and made to be profitable in the market, but as complex ecologies that humans are only a small part of. They continue to treat them as expansive sites of dwelling (Ingold 2000) and urban ecological commons—where they feel entitled to hunt, fish, forage, farm, and build houses, despite not holding property rights.[8]

---

5. See Caroline Merrifield, "Keeping Pace with the Foodshed in Hangzhou," chapter 9 in this volume.
6. See Eric Harms, "Concrete Ecology: Covering and Discovering Saigon's Ecology in a Time of Floods," chapter 8 in this volume.
7. I conducted two months of preliminary research in the summer of 2014 and long-term dissertation research between July 2015 and November 2016. I also completed short-term follow-up research in the summer of 2017 and the spring of 2018. Methodologically, I conducted participant observation and in-depth interviews following the ethnographic tradition in anthropology.
8. While Han Chinese engagement with urban ecological commons is not completely absent in Taipei, it is indigenous engagement with these commons that I focus on in this chapter. This is because indigenous people

In the following, I will first discuss how native people who migrated to the Taipei region flourished in urban ecological commons in the post–World War II period (up to the 1980s). I will focus on urban Pangcah/'Amis communities established on public lands, or *niyaros*, whose living environments were inextricably bound with their surrounding ecologies. I will then show how development projects on Taipei's public lands led by the state and neoliberal capital since the 1990s have resulted in the increasing destruction of urban ecological commons; many indigenous people who were dwelling on the city's riverbanks and hillsides were displaced in this period. In the final part of this chapter, I will show how urban ecological commons have persisted, despite such challenges posed by urban development and landscape transformation, drawing on ethnographic research in two native communities, Kilang and Icep.

The Abundance of Urban Ecological Commons

The Taipei region (*Taibei* in Mandarin) is located in the Taipei Basin, bounded by the Linkou terrace to the west, the Datun Mountains to the north, and the Xueshan Mountain to the south.[9] Because of such biophysical characteristics, Taipei has ample hillside lands (*shanpodi*), which cover nearly 55 percent of the Taipei city and 88 percent of New Taipei City, two major cities in the Taipei region (Soil and Water Conservation Bureau, Council of Agriculture, Executive Yuan 2019). Rivers also figure prominently in Taipei's landscape. The city's major river system is the Tamsui River, which serves as a confluence of three other major rivers: the Xindian River, the Keelung River, and the Dahan River. It was these rivers that were significant to the development of trade and commerce in Taipei's early modern history; the first city center, Mengjia—which means "canoe" in the language of the Ketagalan tribe indigenous to northern Taiwan—was established along the Tamsui River, where native people would engage in trade with the first wave of Chinese settlers (Chiang and Hsiao 1985, 190). Today, rivers continue to play a central role in the provision of drinking water, the maintenance of sewage systems, and, more recently, the development of leisure spaces in the capital region.

Significant urban development in the Taipei region—originally the territory of the indigenous Ketagalan tribe—began in the late Qing period and continued under Japanese rule (1895–1945). But Taipei was a relatively small colonial backwater city until 1949 when the KMT Party (Kuomintang) led by Chiang Kai-shek established a Chinese settler-dominated Republic of China government in exile on Taiwan. The

---

are the ones who have most strongly asserted their right to stay put on public lands and use them as commons, to this day (and to a limited extent, they have gained support from the state). As those whose cultural traditions significantly differ from the Han Chinese, they have been able to articulate their refusal to leave public lands differently from the Han Chinese, who may engage in similar practices.

9. In this chapter, I will romanize Mandarin words using the *pinyin* system. For words of which other romanizations are more common (such as Taipei, Keelung, and Tamsui), I will keep those romanizations.

city's population significantly increased during the KMT authoritarian years, due to massive waves of rural-to-urban migration, as the island industrialized rapidly under KMT developmentalism (Speare 1974). A city with a little over 400,000 people in 1944 at the end of Japanese rule, by 1963 Taipei's population reached 1 million and in 1980 it exceeded 2.2 million (Chiang and Hsiao 1985, 197). Today, the population of the wider Taipei region is approximately 9 million.

Indigenous Austronesian people, the main subjects of this chapter, began to migrate to the metropolis in large numbers under the KMT's industrialization policy during the post–World War II years. Between 1962 and 1971, the indigenous population in western Taiwan, where large cities like Taipei, Kaohsiung, and Taichung are concentrated, saw a nearly 130 percent increase (Wang 1975). In Taipei most native migrants engaged in physically strenuous blue-collar jobs in various sectors such as manufacturing, construction, mining, and fishing (Cai et al. 2001, 21–25). Pangcah/'Amis people were especially prominent in this indigenous rural-to-urban migration due to the relative accessibility of their home villages on Taiwan's eastern coast (Cai et al. 2001, 4). As of 2020, almost a third of nearly 210,000 Pangcah/'Amis people reside in the wider Taipei region (Taipei City, New Taipei City, Keelung City, and Taoyuan City).

The postwar period is often associated with the KMT authoritarian regime's lack of investment in providing "urban commons" for citizens. As far as official institutionalized urban commons like parks and green spaces go, this is true. While the state did have plans to build green spaces and parks for the growing urban population in Taipei, perennial land shortage and haphazard planning meant that most of those plans failed to materialize (Yang and Hirano 1999, 455). However, unofficial urban commons—riverbanks, hillsides, and other gap spaces—were abundant during this time. The general lack of oversight over these commons meant that marginalized people, like indigenous migrants to the city, could appropriate them. My Pangcah/'Amis interlocutors recalled that in newly industrialized and urbanized landscapes within the city they could still find a number of urban ecological commons during the 1970s and 1980s, especially on underutilized public lands near their factories, dormitories, or construction work sites. In these commons, they were able to engage in various indigenous livelihood practices—like fishing and foraging—and even build communities or *niyaros*, to invoke the Pangcah/'Amis concept.[10] One elderly man, Mayaw, who moved to Taipei as a construction worker in 1969, recounted to me how his now forty-year-old community, Kilang, began its life as his fishing spot in the 1970s.[11] Speaking to me on the riverbank of the

---

10. The Pangcah/'Amis concept of *niyaro* is rather difficult to translate. As many anthropologists have shown, it is the basic category of collectivity and social organization in the Pangcah/'Amis social and cultural tradition. According to Michio Suenari (1983), *niyaro* is a microcosm with autonomy, whose main functions are served by its age-set system. *Niyaro*'s publicness is contrasted to the domestic sphere of *loma* (home/house as a built environment) (101). The simple translation of *niyaro* would be "community," but it has the connotation of home (not as a built environment, but as an idea).
11. Kilang is a pseudonym.

Xindian River, where Kilang still stands, he told me that he began to visit that site frequently because it resembled the ecology of his home village in eastern Taiwan. What filled the waterscape at that time were green paddy fields, and none of the high-rises we could see along the river that day existed then. On his visits, he would use a fish net and catch many fish, like carp (which he called *funa* in Japanese)—a practice widely common among Pangcah men at that time in their home villages, according to anthropological research conducted in the 1960s (Qiu 1960).[12] It was not long before he built a hut on the riverbank and began living there, bringing his own family first and then eventually his friends and colleagues. By the 1990s, Kilang had developed into a *niyaro* with over one hundred residents in forty households. Kilang was one of many *niyaro*s that Pangcah/'Amis people established along the Xindian River as well as the Dahan River.[13]

In Icep, another field site of mine, people shared a similar origin story about their community, originally built on a hillside in eastern Taipei. Its founder, Sra, was working as a steel factory worker in Taipei during the 1970s when he stumbled upon the hillside land recently evacuated by the military. During his spare time, he would visit there to pick wild plants—the Pangcah/'Amis are known for extensively using wild, and often extremely bitter, plants in their foodways, even dubbed as a "grass-eating tribe" (*chicao de minzu*) (Sugimoto 2018). Since these plants are not easily available in regular grocery stores, they often forage them in the wild. Like the case of Kilang, the ecology of that hillside public land allowed Sra to maintain practices central to their indigenous cultures. Soon, their friends, relatives, and colleagues purchased (unofficial) land rights from early settlers and built makeshift huts using corrugated sheets and salvaged billboards. By the late 1990s, Icep had become one of the largest *niyaros* that the Pangcah/'Amis ever built in the Taipei region, with over six hundred residents in two hundred households. Without formal access to water or electricity, residents would bathe, wash clothes, and fish in the creek running through Icep. They would use candles or employ generators to get electricity. They would grow and forage plants right next to their huts, and some would also raise chickens, without using cages.

Pangcah/'Amis migrants used urban ecological commons as extensive sites of dwelling. Here I am invoking the concept of dwelling proposed by Tim Ingold (2000). Drawing on Martin Heidegger, Ingold argues that dwelling is any kind of action and movement in relation to the environment. Refusing the distinction between the "built" environment and the "natural" environment, Ingold suggests that human dwelling can happen in relation to many environments. Although

---

12. In the same paper, Qiu (1960) states that while the economy of Fata'an—where many of my interviewees in Taipei came from—was based on rice production at that time, Pangcah men did engage in river fishing during their spare time from agricultural work. Qiu describes elaborate handmade fishing tools that developed as a result of such a fishing culture, such as baskets and fishing nets.
13. I know at least four fully established *niyaros* along these rivers that continue to exist to this day. Many others were demolished, as I will show in the following, or disappeared without leaving many traces but for a few references to them in the newspaper archive.

humans certainly dwell in houses, dwelling can happen in relation to a tree, a river, a forest, and so on. This perspective is helpful here because *niyaros* formed on Taipei's public lands are precisely such sites where the boundary between the "built" part of the "community" and the "natural" environment is unhelpful. Residents dwell in both of them. The occupation of houses is only one of many other dwelling practices that residents engage in, alongside practices like picking snails and fishing.

The Decline of Urban Ecological Commons

Since the 1990s, it has become harder for native people residing in Taipei to sustain urban ecological commons. Public lands that native people have traditionally utilized in the postwar period have been targeted for development by the state and capital. During this period after the collapse of thirty-eight-year KMT authoritarian rule in 1987, the relationship between these two entities became collusive and neoliberal rather than simply top-down as it used to be under authoritarianism. The state and capital—namely big conglomerates and local factions—formed a coalitional relationship in redeveloping and transforming Taipei's urban landscapes (Y.-L. Chen 2005; Hsiao and Liu 1997).

Notable in this new phase of land development has been the prominence of "nature"—not only as an element to be integrated into the urban landscape but also as a form of justifying new urban land developments. The new state-capital coalition no longer sees "nature" simply an object of conquest, unlike at the height of KMT rule; instead, reflecting rising environmental consciousness in this island nation, it has co-opted words like *lü*, "green," *shengtai*, "ecology," and *ziran*, "nature" (Weller 2006; Sun-quan Huang 2012). In this post-authoritarian moment, then, as Arturo Escobar (1996) suggests, the relationship between capital and nature was reimagined as not necessarily oppositional, but a mutually beneficial one; nature was selectively incorporated into its urban development plans—critiquing the lack of parks and green spaces during the authoritarian period and emphasizing the importance of ensuring citizens' access to nature in megacities like Taipei.

However, as Anne Rademacher and K. Sivaramakrishnan point out, "to promote particular urban ecologies may also involve the reproduction or contestation of cultural ideas of belonging to certain social groups, including the city, the nation-state, the region, and the realm called the 'global,' etc." (2017, 12). As we will see below, the creation of middle-class urban natures had destructive consequences; it ended up displacing alternative forms of belonging in Taipei's urban ecological commons.

Among various types of public lands, hillsides (*shanpodi*) and riverbanks (*hebin, he'an*) became perfect targets of development in the Taipei region. Let me first discuss shifts on riverbanks. By the end of the authoritarian period in 1987, Taipei's rivers had become notorious for pollution. There were numbers of garbage dumps on riverbanks, and domestic and industrial waste was directly dumped into rivers with no effective regulation to speak of (Chi 1987). Damages from frequent floods also constituted a major issue in urban governance. Managing such polluted and unruly rivers and transforming them into safe middle-class spaces of leisure became central to the state's riparian policies during the democratization period after 1987.

I will particularly zoom in on Taipei County, where many indigenous communities existed along its rivers. You Qing, the Taipei County magistrate who served between 1989 and 1997, made it one of his central policies to clean up trash accumulating along the rivers running through his county. By the late 1980s, without an effective waste disposal system, thirteen enormous garbage dumps (*leseshan*) had formed on riverbanks, located throughout the county (You, Sun, and Lin 2017, 71). You Qing spearheaded the projects to remove these garbage dumps from riparian locations; after eliminating them, he proposed to turn riverbanks into sites of middle-class leisure. One planning document published by the Taipei County government in 1993 states: "Tourism and recreation will put at its center servicing urbanites and prioritize the development of community parks, high riverbanks (*gaotandi*) in river reservation areas, and water channels" (Taibeixian zonghe fazhan jihua 1993, 48). The document encourages the Water Management Department of the county to develop high riverbanks of the Xindian River and the Dahan River so that urban residents can use them for recreational purposes (Taibeixian zonghe fazhan jihua 1993, 48). These visions did materialize over time, and the county government gradually covered the banks of the Dahan River, the Xindian River, and the Keelung River with dykes, riverside parks, and various environmental amenities—bicycle paths as well as basketball, badminton, and tennis courts. Private capital has joined this landscape transformation, developing waterfront housing complexes (*shui'an zhuzhai*) near these tamed riverscapes, which have become popular among upper-class homebuyers (Wang and Lee 2015). Once despised for their smells from industrial and domestic waste flowing in them, the Taipei County's riverscapes are today bustling with retirees walking their dogs, high school students playing basketball, and joggers in fashionable running gear (Figure 6.2).

The landscape transformation has been no less dramatic on hillside lands. Although approximately 60 percent of the Taipei region is not suited for urban development because of its slope, the rapid post–World War II population growth in the capital and the speculation boom means that development, rather than sensible planning, has been prioritized (Selya 1995, 163–64). Real estate developers had begun to purchase hillside lands as early as the 1970s, exploiting loose development regulations; despite the introduction of stricter regulations on hillside development in the early 1980s, developers claimed that old permits issued before that were still valid and built a number of housing complexes on unstable hillside lands (Hsiao and Liu 2002, 72). In his well-studied account of the real estate development frenzy in Taipei since the 1980s, Chen Dong-sheng notes toward the second half of the 1980s, especially after the real estate bubble in 1987, middle-class Taipei citizens

Figure 6.2: A riverside park in New Taipei City

were in search of more affordable housing, and hence found such suburban complexes built on hillsides to be attractive options (Chen 1995, 231–32). Proximity to nature and distance from the bustling, polluted central city became an important selling point for these housing communities, which were advertised as *gaoji bieshu* ("upscale villa"). As a result, hillside development was extremely lucrative. One housing community, Longshanlin—built in 1987 on a hillside located not far from Icep—garnered the pure profits of 4 billion New Taiwanese Dollars (Chen 1995, 247). Many developers that developed hillsides were major ones, like those from the well-known conglomerate called the "Gang of Sanchong" (*Sanchong bang*) (Chen 1995, 253).

It is clear from these cases of development on hillsides and riverbanks that certain forms of nature have come to be fetishized and commodified in post-authoritarian Taipei. The impact of such urban development has not been small on native migrants who were informally using Taipei's hillsides and riverbanks as their urban ecological commons, especially by building collective sites of dwelling, *niyaros*. Many *niyaros* have been subject to violent displacement as a direct result of landscape transformation on both riverbanks and hillsides. In December 1995, on the right bank of the Xindian River being redeveloped by the county government, one *niyaro*, called Xiaobitan, was demolished. One Xiaobitan resident angrily said to a reporter: "They said they would not demolish our community until we had found somewhere else to live. They broke their promise."[14] Mayaw Biho's 1997 documentary *Children in Heaven* (*tiantang xiaohai*) records the multiple displacements of San'ying, a small *niyaro* built along the Dahan River, between 1996 and 1997. Icep, one of my field sites, was relocated in 2000, because the state planned to use the hillside it was occupying as a site for building a high-speed rail facility. More recently, San'ying, which was rebuilt on the Dahan riverbank after the demolitions in the 1990s, was demolished again in early 2008 for a riverside park project (which was captured again by Mayaw Biho in his subsequent 2010 film, *A Big River in Front of My House* or *wojia mianqian you dahe*). Several kilometers south of San'ying, also along the Dahan River, the Taoyuan County government announced a plan to demolish Saowac for the construction of a bicycle path on the riverbank. Despite vociferous protests by residents and their supporters, on February 20, 2009, the police brought excavators and destroyed this *niyaro* (Chen and Qiu 2014).

However, the state-capital coalition has been unsuccessful in completely taking over Taipei's hillsides, riverbanks, and other public lands. Ecological urban commons persist. They are latent, to invoke Anna Tsing again—they "move in law's interstices" (2015, 255). By staying under the radar, they "bubble with unrealized possibilities" (Tsing 2015, 255). My Pangcah/'Amis interlocutors in Taipei continue to gather wild plants and open up gardens, catch snails, and even hunt animals like boars in such latent urban ecological commons. They find and form these commons in awkward spaces under freeway overpasses; on abandoned hillsides where real estate development is not possible; and small edge spaces on riverbanks adjacent to brand-new bicycle paths. They defy the enclosure and commodification of hillsides and riverbanks. It is to these ecological urban commons that I will now turn.

Urban Ecological Commons 1

Kilang stands along the Xindian River, located in the southern part of the Taipei region. Since a Pangcah man, Mayaw, settled here in the 1970s, Kilang steadily grew and became a *niyaro*. Today over a hundred Pangcah/'Amis people—across generations from small infants to high-school students to retirees—call this community home, and the *niyaro* has over forty houses, karaoke bars, corner shops, and eateries. In the community's center there is a large plaza facing the Xindian River. It serves as an important common space among residents, where they can hang out with neighbors and friends, over drinks, cigarettes, and betel nuts. It is also where important cultural rituals and events are held, like *ilisin*, the annual harvest festival important in the Pangcah/'Amis tradition.

The Xindian River has always functioned as an important urban ecological commons among Kilang residents. I already mentioned Mayaw's fishing story earlier in this chapter. Another long-time resident, Ichang, talked about how he

---

14. I thank Cao Wen-jie for sharing a news clip about Xiaobitan's demolition with me.

used to bathe in the Xindian River. Pointing to the river, he said, "I would go over there, use a little bit of soap, wash my hair, and come back." He added: "The Xindian River was clean before. There used to be a lot of shells. We could eat them" (Personal communication, 2016). Other residents told me that they could even swim across to the other side of the river before. But Kilang residents mostly used the past tense for these recollections—they are acutely aware that the Xindian River and its surrounding ecologies have changed dramatically over the last few decades; they sense that their urban ecological commons have shrunk in size and scale.

This transformation began especially after the Taipei County government launched the Rhine Project in 1993. Named (of course) after the well-known Western European river, the Rhine Project was meant to develop particularly the right bank of the Xindian River for the purpose of tourism and recreation, while preserving the water quality upstream. Following this project, in 2007, the Taipei County government, under the new magistrate Zhou Xi-wei, announced yet another plan to create more leisure sites like riverside parks along the Xindian River, this time on the left bank, where Kilang is located. That year, the county government proposed to relocate Kilang to public housing.

Kilang residents challenged this relocation plan. They demanded the right to dwell along the Xindian River. They put up banners throughout their community. They went on protests in front of the county government. This political action was featured prominently on national media, bringing light to the predicament of urban indigenous people—newly recognized as one of the most marginalized groups in Taiwan following the collapse of KMT rule. This activism paid off, and luckily, unlike many *niyaros* I briefly discussed above—like San'ying and Saowac, which were demolished in 2008 and 2009, respectively—Kilang was narrowly able to avoid demolition and relocation. While Kilang was not demolished, the county government went forward with the construction of riverside parks and bicycle paths on the left bank. These environmental amenities were built adjacent to Kilang, almost directly passing through this small indigenous community. Hence, today two kinds of urban commons uneasily coexist along the Xindian River: on the one hand, there are official urban commons, or "public goods," as Gidwani and Baviskar put it (2011, 43)—benches, gazebos, and bicycle paths built and maintained by the state. On the other hand, Kilang residents continue to treat the Xindian River and its surrounding ecologies as urban ecological commons, as the less institutionalized, latent kind that I have been following. They continue to engage in fishing, foraging, and gardening, despite inconveniences imposed by park construction and the official, formalized commons.

Spending time with Kilang residents, I was constantly reminded of the persistence of urban ecological commons, both as an ideal and an actual reality. On one April day in 2016, I was invited to visit the house of Kufi, a Pangcah woman who was taking my weekly Japanese class in Kilang. As we ate dinner watching TV together, Kufi's husband served me fish soup and explained that he caught the tilapia fish (*wuguoyu*) in the soup from the river "down there" (*xiamian zhuade*). While saying that, he pointed to the Xindian River running behind their hut. Some residents do complain about the pollution in the river and the smell that fish possess, even after its water quality supposedly improved in recent years. But many people, especially men like Kufi's husband, continue to catch fish in the Xindian River, and fish from there regularly end up on residents' dinner tables.

The *niyaro*'s surrounding ecologies also provide ample foraging sites. On a rare cold winter day in Taiwan in early 2016, I paid a visit to Kilang again, and Banay—a middle-aged woman who is the daughter of the founder Mayaw—was outside, warming herself around a fire. There, she was picking leaves from the stems of *tatokem* (*Solanum nigrum*, blacknight shade) and *samah* (*Lactuca indica*, Indian lettuce), two very common wild and bitter plants in Taiwan commonly eaten by the Pangcah/'Amis. Banay said that she went to the mountain "back there" (*houmian*) and came back with a box full of these wild plants. If Kilang residents had relocated, they would have lost access to these important foraging sites and ways to access food items that are important to them. After I finished teaching my Japanese class there, Banay sent me home with a bag of *tatokem*.

Kilang is also filled with gardens. Even after the construction of parks and bicycle paths around them, residents have continued to use spaces throughout their community for growing various plants like taro, shiso leaves, and bitter melons. To the untrained eye, these may seem like a collection of weeds. But small spaces situated awkwardly next to a street and electric poles can surprisingly nurture a wide range of plants. Along the brand-new riverside park and bicycle path built by the government, Kilang residents remind us that the entanglements of humans and rivers different from those envisioned by the state continue to exist. Two different visions of urban commons are pushing against each other here.

Urban Ecological Commons 2

I will now shift to urban ecological commons in a different Taipei neighborhood. Formerly a *niyaro* located on a hillside in eastern Taipei, Icep is today housed in three nondescript mid-rises that house approximately one hundred and twenty households. This relocation caused massive opposition from residents, who went on various protests. However, unlike Kilang, they were not able to avoid the relocation. Approximately half of the residents living there are Pangcah/'Amis people relocated two decades ago, while the other half are new residents who moved in after original relocatees moved out. The housing complex is located in a typical urban neighborhood in Taiwan, surrounded by other high-rises, convenience stores, breakfast shops, and small eateries. It seems that Icep residents have lost access to urban ecological commons. However, some continue to seek out uninstitutionalized, latent urban ecological commons in this urban neighborhood.

The land where Palafang owns her garden—discussed in this chapter's introduction—can be characterized as one such commons. That is actually located on the hillside where Icep stood until 2000. The state's plan to build a high-speed facility never materialized, and people like Palafang began to return there. As I said that hillside today is filled with a number of gardens, and plants and vegetables nurtured there are sold on the street to Icep residents.

I have come to know several more urban ecological commons over the course of my ethnographic research with Icep residents. One winter day in 2015, my friend Sawmah—a Pangcah woman in her early seventies who lives in Icep, works as a janitor, and also owns a casual eatery next to Icep—offered to take me to one of them. While we were chatting at her eatery, Sawmah explained that she knew a *dongec* garden not far from us. Rattan's stem, or *dongec* as the Pangcah/'Amis call it, is often cooked in soup with pork spareribs. While it is extremely bitter, Pangcah/'Amis enjoy and crave it precisely for its bitterness. Many Pangcah/'Amis elders do not miss a chance to forage *dongec* in the wild or domesticate it in gardens. After Icep was relocated in 2000, Sra, the now-deceased founder and former chief of Icep, found this spot—just like he stumbled upon the former territory of Icep—and planted some *dongec* there. After Sra died about a decade ago, Sra's widowed wife continued to take care of the garden. But now she is too old and sick to maintain this garden any longer, having suffered a stroke several years ago. So she asked Sawmah's still-healthy husband, A-kuan, a Pangcah man in his eighties, to tend to it.

Soon after leaving Sawmah's eatery, we ventured off the neighborhood's main road. Motorcycles, taxis, and buses were no longer driving past us. We didn't encounter any other pedestrians. Now in the part of the neighborhood I had little familiarity with, I followed Sawmah. We soon began seeing cars parked and abandoned on both sides of the road. After traversing several concrete freeway overpasses, we arrived at a large parking lot. Sawmah explained that the driving school that used to be there went out of business several years before. A Han Chinese family who owned a house there has moved out as well. Having lived in this neighborhood for the last four decades, she is intimately familiar with all the changes that have happened there. Next to the abandoned school, what used to be its parking lot, and an empty house—stray dogs were roaming around in the lot—a thick foliage of trees spread. We continued our walk in that direction.

After several minutes of walking, we reached the end of the dirt road and arrived at Sra's *dongec* garden. Indeed, there were rattan palms planted by Sra when he was still alive (Figure 6.3). I was quite surprised that *dongec* had a number of spikes on its surface; they need to be removed to acquire the softer, yellow stem eaten by Pangcah/'Amis people. Not unlike Palafang's plot I discussed in the introduction, this garden was also in such an awkward location—on an inconveniently located land that cannot be possibly developed into anything. Only those who are interested in utilizing its ecologies would even think about settling on a piece of land like this.

Figure 6.3: Sra's rattan garden

But for Sra, this was a perfect urban ecological commons. Further inside the garden, he had built a large wooden hut. Thick comforters folded inside it suggested that he used to sleep there. Next to the hut was a small Catholic church—due to intense missionary work in indigenous communities during the post–World War II period, many native people in Taiwan are Christian (Huang 1996). An old mildewed copy of the Bible written in the Pangcah/'Amis language was sitting on a night table, next to reading glasses. Sawmah explained to me that the Virgin Mary statue sitting inside used to be in the Catholic church at Icep, which was, like all other buildings, demolished at the time of the 2000 relocation. Sra kept the statue and when he rebuilt the church next to his clandestine *dongec* garden, he placed it there instead.

Sawmah and I returned to the same garden a few years later, in March 2018. When we reached Sra's garden, I was shocked that everything had changed completely. It was clear that Sawmah's husband, A-kuan, had stopped attending to it. The wooden walls of both the hut and church had rotted. Rattan trees, which had looked so vibrant only a few years before, were also wilting. While Sra's garden had been abandoned, there were signs that this land was still functioning as urban ecological commons among Pangcah/'Amis people living nearby; other Icep residents had set up gardens on this land. As we finished looking at Sra's garden and began to retrace our steps, we ran into Sawmah's neighbor from Icep, Nakaw. A Pangcah woman in her seventies like Sawmah, Nakaw came out of a small garden of her own. We asked her what she grew, and she responded, "Not that much. Just some *lakiu*"—scallions often eaten fresh with salt by Pangcah/'Amis people. While saying that, she found some *pahko*—a type of fern (*Diplazium esculentum*)—growing on the ground and picked up some for us to take home. Sawmah found what Pangcah/'Amis call *halufaw*, water celery growing in the wild, so we picked it together as well. Lands like this are hence important multispecies assemblages—where Pangcah/'Amis people can derive nurturance in the city. They mostly escape the notice of urban dwellers. But they lay dormant and latent, inviting both humans and nonhumans to nurture them.

Coda

What is the temporality of these urban ecological commons? Do they ever come to an end? Do they ever last? During the 1970s and 1980s, Pangcah/'Amis people could still find belonging in Taipei's urban ecological commons. They could more or less freely fish, forage, garden, hunt, and even build large-scale *niyaros* on the city's public lands, without inviting that much unwanted attention or policing from the state and/or landowners. Since the 1990s, the temporality of urban ecological commons has become decidedly shorter. The Taiwanese state, in collaboration with capital, has adopted new imaginaries of urban natures; they have transformed Taipei's riverbanks, hillsides, and other public lands into spaces of leisure and sites of middle-class living. From the perspective of native people, material and discursive transformations that these lands have undergone were profoundly unsettling— leading to the destruction of their communities and their displacement from urban ecological commons, sites that they have long utilized for fishing, foraging, hunting, gardening, and other dwelling practices. Today, Pangcah/'Amis people are forced to inhabit what Erik Harms calls "a liminal state of ruptured time" (2013, 345) in urban ecological commons, as they constantly worry that they may be taken away.

However, as I have shown, ecological urban commons have persisted. The forced relocation of people living in these commons has certainly caused overt political actions like protests, as we saw in Kilang and Icep. However, their political potential lies also in being "latent" (Tsing 2015, 255)—hidden, lying dormant, and precisely because of their latency, filled with possibilities, at least before they become targets of development. By enacting distinct human-environmental relations in these commons, indigenous people in Taipei challenge the state-capital alliance's visions of how nature should be nurtured on Taipei's public lands. They critique how official green urbanist initiatives have disabled so diverse ways of engaging with Taipei's rivers, hillsides, public lands, and a wide range of more-than-human-beings that flourish there. They show that biking, jogging, walking dogs, and playing sports inside riverside parks or living in exclusive hillside housing communities are not the only ways to live with urban natures. By maintaining and thriving in urban ecological commons, they unsettle the logics of determining what kinds of natures should be allowed to live, and what kinds should be let die.

Works Cited

Bayat, Asef. 2013. *Life as Politics: How Ordinary People Change the Middle East*. Stanford, CA: Stanford University Press.

Cai, Ming-zhe et al. 2001. *Taiwan yuanzhumin shi: dushi yuanzhumin shipian* [The history of indigenous Taiwan: Volume on urban indigenous people]. Nantou: Guoshiguan Taiwan wenxianguan.

Chen, Dong-sheng. 1995. *Jinquan chengshi* [Plutocratic city]. Taipei: Juliu.

Chen, Yi-Ling. 2005. "Provision for Collective Consumption: Housing Production under Neoliberalism." In *Globalizing Taipei: The Political Economy of Spatial Development*, edited by Reginald Kwok, 99–119. New York: Taylor and Francis.

Chen, Yong-rong, and Yan-liang Qiu. 2014. *Fang tianzai yu renhuo: yuanzhumin kangzheng yu Taiwan chulu* [Against natural and human disasters: Indigenous protests and Taiwan's future]. Taipei: Taiwan shehui yanjiu zazhishe.

Chi, Zong-xian. 1987. *Taiwan de xuemai: women de hechuan xunli* [Taiwan's blood vessels: Touring our rivers]. Taipei: Lianhe yuekan zazhi she.

Chiang, Nora, and Michael Hsin-Huang Hsiao. 1985. "Taibei: History and Problems of Development." In *Chinese Cities: The Growth of the Metropolis since 1949*, edited by Victor Sit, 188–209. Oxford: Oxford University Press.

Escobar, Arturo. 1996. "Construction Nature: Elements for a Post-Structuralist Political Ecology." *Futures* 28 (4): 325–43.

Escobar, Arturo, Dianne Rocheleau, and Smitu Kothari. 2002. "Environmental Social Movements and the Politics of Place." *Development* 45 (1): 28–36.

Gidwani, Vinay, and Amita Baviskar. 2011. "Urban Commons." *Economic and Political Weekly* 46 (50): 42–43.

Harms, Erik. 2013. "Eviction Time in the New Saigon: Temporalities of Displacement in the Rubble of Development." *Cultural Anthropology* 28 (2): 344–68.

Holston, James. 2019. "Metropolitan Rebellions and the Politics of Commoning the City." *Anthropological Theory* 19 (1): 120–42.

Hsiao, Michael Hsin-Huang, and Hwajen Liu. 1997. "Land-Housing Problems and the Limits of the Non-Homeowners Movement in Taiwan." *Chinese Sociology & Anthropology* 29 (4): 42–65.

Hsiao, Michael Hsin-Huang, and Hwajen Liu. 2002. "Collective Action toward a Sustainable City: Citizens' Movements and Environmental Politics in Taipei." In *Livable Cities? Urban Struggles for Livelihood and Sustainability*, edited by P. Evans, 67–94. Berkeley: University of California Press.

Huang, Shiun-wey. 1996. "The Politics of Conversion: The Case of an Aboriginal Formosan Village." *Anthropos* 91: 425–39.

Huang, Sun-quan. 2012. *Lüse tuituji: jiuling niandai taibei de weijian, gongyuan ziranfang dichan yu zhiduhua dijing* [Green bulldozer: Squatters, parks, ecological real estate, and institutionalized landscapes in 1990s Taipei]. Taipei: Pozhoubao.

Ingold, Tim. 2000. *The Perception of the Environment: Essays on Livelihood, Dwelling and Skill*. New York: Routledge.

Qiu, Qi-qian. 1960. "Matai'an ameizu de yulao shenghuo" [The fishing life of the Fata'an Pangcah]. *Zhongyang yanjiuyuan minzuxue yanjiusuo jikan* 10: 57–84.

Rademacher, Anne., and K. Sivaramakrishnan. 2017. "Introduction: Places of Nature in Asian Cities and Towns." In *Places of Nature in Ecologies of Urbanism*, edited by Anne Rademacher and K. Sivaramakrishnan, 1–26. Hong Kong: Hong Kong University Press.

Selya, Roger. 1995. *Taipei*. Chichester: John Wiley and Sons.

Soil and Water Conservation Bureau, Council of Agriculture, Executive Yuan. 2019. "Shanpodi ziben ziliao shuoming." Accessed July 15, 2019. https://www.swcb.gov.tw/Topic/show_detail?id=4d4cf309c3f3424786b765c3cc6383ec.

Speare, Alden Jr. 1974. "Urbanization and Migration in Taiwan." *Economic Development and Cultural Change* 22 (2): 302–19.

Suenari, Michio. 1983. *Taiwan ami zoku no shakai soshiki to henka: mukoirikon kara yomeirikon he* [The social organization and change of the Pangcah/'Amis Tribe in Taiwan: From uxorilocal marriage to virilocal marriage]. Tokyo: University of Tokyo Press.

Sugimoto, Tomonori. 2018. "'Someone Else's Land Is Our Garden!': Risky Labor in Taipei's Indigenous Food Boom." *Gastronomica: The Journal of Critical Food Studies* 18 (2): 46–58.

Susser, Ida. 2017. "Commoning in New York City, Barcelona, and Paris." *Focaal: Journal of Global and Historical Anthropology* 79: 6–22.

Taibeixian zonghe fazhan jihua. 1993. *Bumen fazhan jihua (3): guanguang xiuxian, laogong, shehuifuli, weisheng yiliao* [Departmental development plan no. 3: Tourism/leisure, labor, social welfare, and hygiene/medicine]. Taipei: Taibei xian zhengfu.

Tsing, Anna. 2015. *The Mushroom at the End of the World: On the Possibility of Life in Capitalist Ruins*. Princeton, NJ: Princeton University Press.

Wang, Chih-Hung, and Han-Ru Lee. 2015. "Lüse jinshenhua? Taibei duhuiqu shuian zhuzhai fazhan chutan" [Green gentrification? Preliminary research on the development of waterfront housing in the Taipei metropolitan region]. *Shehui kexue luncong* 9 (2): 31–88.

Wang, I-shou. 1975. "Mountain People in the Lowlands: A Preliminary Report on the Migration of Formosan Aborigines." *Proceedings of the Association of American Geographers* 7: 264–68.

Weller, Robert. 2006. *Discovering Nature: Globalization and Environmental Culture in China and Taiwan*. Cambridge: Cambridge University Press.

Yang, Pin-an, and Kanzo Hirano. 1999. "Taipeishi no kōen ryokuchi no hensen to kōsatsu" [Research on the transition of city parks and open space in Taipei City]. *Landscape Research Japan* 62 (5): 453–58.

You, Qing, Yan-long Sun, and Feng-bin Lin. 2017. *Juanjuan huaigu: taibeixian zhengfu de naxienian* [Reflecting on memories: Those years at the Taipei County government]. Taipei: Xiuwei.

7
Making Land Out of Water: Ecologies of Urbanism, Property, and Loss

Shubhra Gururani

In July 2016, the rains played havoc and brought the city of Gurgaon to a complete halt. Within a few hours of the rain, one of the main drains—the Badshahpur *nallah* that runs through the city—breached and flooded the Delhi-Gurgaon expressway. Thousands of commuters ply the expressway from Delhi to Gurgaon and back each day, but that evening hundreds of vehicles and commuters were stuck in the worst-ever traffic snarl that the city had seen. The backup extended fifteen to twenty kilometers, by some accounts. It was reported that many of those caught in the traffic jam, or what soon came to be called the "Gurujam," were forced to spend the night inside their vehicles, while others just abandoned their vehicles and waded through the water to make it home. As countless images and Twitter feeds of submerged cars and trucks flooded the news and social media, schools were shut down and the district magistrate passed emergency orders to manage the chaos. The flooding and the jam, as one reporter noted, was an "unfolding urban nightmare" (Choudhry 2016).[1]

Gurgaon, recently renamed Gurugram, is located at the southwestern edge of New Delhi and has come to stand for India's poster child of real estate boom, "high-end" urban living, and for privatized land and housing development more broadly. Gurgaon's terrain is rocky, sedimented by the red Aravalli mountain range that extends from the neighboring state of Rajasthan and continues into Delhi. Given its mountainous topography, Gurgaon is historically associated with drought-like conditions and a depleting water table, leading to water shortages (see Narain and Singh 2017). In the colonial archives, there is considerable discussion of the brackish quality of water and poor drainage that negatively impacted agricultural production and livelihoods in this southern tip of colonial Punjab (Government of Punjab 1883–1884, 5; Thorner 1995, 66). In the 1880s, when the irrigation network in Punjab was vastly extended by the British,[2] the hydraulic engineers deemed it to

---

1. Since then, there have been more major incidents of flooding and traffic jams.
2. In the 1880s, with nine major canals colonial Punjab had one of the largest irrigation systems in the world (Agnihotri 1996). Interestingly, while the British between 1898 and 1948 promoted canal irrigation in Central and West Punjab, they actively discouraged irrigation and agriculture in the arid region of East Punjab where

be too costly to stretch the canal network to the southern edge of the district where Gurgaon is located. As a result, Gurgaon has remained a mostly waterless terrain that many longtime residents, like Ragbar, a seventy-year-old Gujjar male, described as "just rocks and dirt! . . . There was no water here and we could barely grow one crop, and if the rains failed, then [there was] nothing!" (personal communication, December 18, 2018). After India's independence in 1947, when Delhi's first Master Plan was drafted in 1962 Gurgaon was identified as one of the five "ring towns" that could potentially accommodate future urban expansion. However, it was exempted from planned urban development as it was deemed to be in an earthquake-prone zone and was considered "handicapped" due to lack of good water resources (see Gururani 2013). While the neighboring ring towns of Faridabad and Ghaziabad started to urbanize, Gurgaon until recently remained a cluster of agro-pastoral villages.

There are no perennial rivers running through the district of Gurgaon and the sole river, the Sahibi, also sometimes called Sabi, found mention in colonial gazetteers only because it was whimsical and its flow could not be relied on. Interestingly, despite the aridity of the region, floods were not unknown and the district gazetteer noted that "In a normal year the lands are not flooded. . . . It is only when the flood comes down that it sweeps over the country destroying the crops" (Government of Punjab 1883–1884, 4). The local residents today also vividly recalled the floods of 1964 and 1967, but it was the floods of 1977 that were widely considered to be a game-changer for New Delhi and, with that, also for Gurgaon.[3] Around the same time, the state of Haryana was carved out of Punjab in 1966 and the first private land developers, like the DLF, started to buy agricultural land in the 1970s and convert it to urban real estate (Gururani 2013). Amid rapid conversion of agricultural land into real estate, environmentalists and local citizen groups have raised alarm and drawn attention to the ecological challenges posed by such extensive urbanization (Sohail-Hashmi 2016). The Centre for Science and Environment recently noted that Gurgaon is faced with the twin challenges of floods and droughts (Roychowdhury and Puri 2017). Yet, urbanization of agrarian hinterlands has continued unabated.

The editors of this volume have invited us to reflect on how the social and biophysical entanglements of urban spaces are being (re)articulated in this moment of unprecedented urban transformation. They asked us to consider how cities and their futures are being imagined and inscribed, while remaining attentive to the pluralities and potentialities these reconfigurations may offer. In thinking along

---

Gurgaon is located. This arid terrain was left for raising a supply of draft animals and for cultivating low-value food-cum-fodder crops, and carved out as a "backward region" (Chowdhury 1986, 263).

3. In 1961–1962, to mitigate the possibility of flooding Delhi, the Municipal Corporation of Delhi built a check dam at the then-border of Delhi and Haryana, where the Najafgarh Nallah enters Delhi. But, soon after the dam was built in 1964, as luck would have it, the Yamuna flooded and the Sahibi and Najafgarh Lake were identified as the two main culprits. Subsequently, recommendations were made to widen the Najafgarh Nallah to regulate the water before it entered Delhi and impacted the flow of water in Haryana and Rajasthan.

these lines, this chapter takes the flooding of Gurgaon in 2016[4] as a point of entry to reflect on the changing ecologies of urbanism and to situate flooding in the deeply political processes in which an alliance between the state and politicians, agrarian elites, land developers, brokers, and bureaucrats is implicated in orchestrating the death and life of urban nature. In the context of climate change, with warming temperatures and rising sea levels, there is a growing recognition that coastal cities are faced with severe ecological challenges, as contributions to this volume show. But it is important to note that the flooding of Gurgaon was not due to rising water levels of rivers or lakes: instead it registers a different kind of interplay, a political-ecological ruination that can be described as what Veena Das in another context has called a "critical event." According to Das, a critical event conveys "*a new modality of historical action* which was not inscribed in the inventory of that situation" (Das 1995, 5, emphasis in original).

I argue that flooding in Gurgaon marks a critical *political-ecological* event that brings together a range of human and nonhuman actors and alerts us to the rupture or even collapse of sedimented ecological processes and everyday practices of habitation. It offers an opportunity to trace the material and discursive ways through which the rhythms of space and time are realigned and to evaluate how historically situated boundaries between land and water, society and nature, and rural and urban are recalibrated, flattened, and reordered. It also offers a window into how new imaginaries of the urban are enabled to accommodate dreams and desires in the context of what Goldman has described as "speculative urbanism" (Goldman 2011). Against this backdrop, tracking the course of the floods necessitates asking how the vast hydrological network was transformed, how water bodies were made to disappear, how existing regimes of property, tenure, and livelihood changed, and importantly, how new meanings and values scripted to assetize not only land but also, as I argue below, water.

The alterations in the rhythms of land and water that I describe are by no means new to the region nor are they unique to Gurgaon (see Sengupta 2016). The disappearance of lakes in Bangalore (see Ranganathan 2015), the reclamation of coastal land in Mumbai, the making of Calcutta (see Bhattacharyya 2018), or the cases of Saigon in this volume[5] are some of the many examples that tell a compelling story of how the boundaries between land and water are blurred or sometimes ignored, to pave the way for geographies of the urban to take shape. And these stories of urban ecologies, as Rademacher notes, always unravel how and why "certain ecological logics [a]re made legible, powerful, and active" and how local relations of power articulate with the unfolding geography of value, land, property-making, and speculation (Rademacher 2011, 176; see also Bhattacharyya 2019). In Gurgaon District, where the city of Gurgaon is located, by some accounts as many as 389

---

4. Since 2016, Gurgaon has witnessed flooding annually, most recently in August 2020.
5. See Erik Harms, "Concrete Ecology: Covering and Discovering Saigon's Ecology in a Time of Floods," chapter 8 in this volume.

bodies of water have disappeared in the past sixty years (Arora 2018), of which 137, including some fairly old lakes, have dried up in the last ten years to make room for the so-called India's Millennium City (Kumar 2012, updated 2017). These water bodies were crucial to maintaining the regional hydrology that would seasonally enliven the arid landscape, saturate the parched lakebeds, and prepare the soil for crops, vegetables, and pasture for local residents to make a living. They were also crucial for channeling rainwater into the Najafgarh Basin and prevent flooding. But, since the mid-1980s and the liberalization of India's economy, the financialization of land has meant that the networks of drains, ditches, wells, canals, ponds, and creeks have been covered up, blocked, and virtually erased for the city yet to come. Not only were hundreds of water bodies literally concretized, remade into land assets, but many have also disappeared from government maps and plans, and certainly from the gaze of most private developers and new residents who have recently come to live and work in Gurgaon.

One lake—Ghata Jheel (or Ghata Lake), which I discuss below—is one of several bodies of water that have disappeared. The Ghata Jheel was an old lake, one of the largest and deepest water bodies in the village of the same name, Ghata, located in what is now the city of Gurgaon. The Jheel was described in the 1910 District Gazetteer of Gurgaon as part of the North Gurgaon drainage system. It had been carefully secured by an embankment (*bundh*) that was built around 1889. The term, Ghata, in Hindi and Haryanvi (the local dialect) refers to a gap or a low-lying area and also to loss. In Gurgaon, it cartographically refers to the 1.5 kilometer-long stretch between two villages—Ghata and Behrampur and served as a conduit for carrying rainwater from the Kaderpur-Mehrauli ridge to the Badshahpur drain and into the Najafgarh drain/basin, which is the only waterway to drain water into the Yamuna River to the east. KS,[6] a Gujjar elder sitting in front of his small grocery store at the edge of the dried-up Jheel in 2018, pointed his finger and explained:

> This entire region is very uneven, rocky. It is full of deep ditches, craters, and trenches.... There are five small *jhors* (ponds) in the village and then there was the Ghata *jheel*. When it rained the entire area used to fill up with water. The British built a dam more than a hundred years back; it is still there and until recently it was regularly maintained. There was a passage (*mori*) for the water, which was closed in June to collect all the rainwater, and then drained after Diwali in October... water came down from Mehrauli to Mandi Gaon to Gwal Pahari to Bandhwari and then to Ghata. In the month of *Kwar* (September), when the jheel would overflow, we let the water enter into our fields. The water would run deep and wet the soil, and on it we would plant *desi gehu* (local wheat) which would grow this high, almost as tall as a man. We would then harvest and eat that wheat. The rains would feed the water level of the jheel and we would dig holes in our land and they would fill up with water which we would keep for drinking. Similarly, we would dig holes for our

---

6. To maintain anonymity, I have either changed the names or used initials of the individual I spoke with during fieldwork.

animals as well; there were separate ditches for our animals and us near the bundh down there... There was a lot of rain in 1972 and the lake filled up in 1977, but since then the rains have diminished and it has barely filled up. Now, you can go even two hundred feet [deep] there is nothing, borewells have sucked all the water. (Personal communication, December 2018)

The area of the lake is estimated to have extended to close to 300 acres but it has a vast catchment area upstream of 33,000 acres, of which 10,000 acres lie in the eight villages in Haryana and the rest lie in Delhi. In the wake of extensive urban development, even though the Ghata Lake exists in the minds and memories of local rural residents, it has almost disappeared from official maps and plans. A recent report on the state of water bodies in Gurgaon noted, "there is no government/public owned waterbody upstream of the Ghata bundh [embankment] which could be defined as 'Ghata Jheel or Ghata Lake'. The Ghata bundh itself now survives as a 800 m long structure" (Gurugram Metropolitan Development Authority 2019, 15).

The defacing of Ghata and other water bodies can be located at the intersection of overlapping processes of property making, urban planning, speculation, and ecological restoration. Embedded in the anticipatory telos of an imminent urban future, speculation works to bolster the aspirations of a growing middle class and conjures dreams of modern living, homeownership, luxury, and a range of new aesthetics. It relies on discursive strategies that often obfuscate or manipulate documents, plans, memories, and everyday practices. It also often recruits a powerful ecological imaginary, as it did in Gurgaon, of lush landscapes, sustainability, and natural heritage, an imaginary that not surprisingly values landed property over watery ones. In this conjuncture, amidst extensive speculation all watery bodies, ponds, ridges, ditches, depressions, and wastelands have come to be seen as what Bhattacharyya has described, in another context, as "land-in-waiting for property development" (Bhattacharyya 2018, 169): a watery terrain that can be and should be solidified and assetized.

Turning ponds and lakes into land is a convoluted process. It is a process that is not just about reckless urban development, which it is to some extent, or an act of greed or profiteering, which it is too. The act of making new urban ecologies first and foremost entails erasing boundaries, especially but not only between land and water, and reconfiguring existing regimes and practices of property. It entails manipulating local histories of access and authority, and gradually introducing a new lexicon of financialization and assetization of land. It engenders the will to embrace risk and uncertainty and involves a range of practices that, Bear et al. suggest, "produce an instability of value" (Bear, Birla, and Puri 2015, 387). The "instability of value" is pivotal as it not only introduces new registers of value, aesthetics, and sociality, but it also generates a productive obscurity and paves the way for new social-ecological relations, and with them new social and ecological imaginaries. The retelling of stories of land, water, and property, and their manipulation works to obscure situated (agrarian) ecologies and the associated everyday practices of land, property, and livelihood. In place of agrarian practices, it promotes a hegemonic vision of nature that Raymond Williams calls a "pleasing prospect"; a prospect of a clean, green, and investible periphery (Williams 1973; see also Mozingo 2011). But, as Rademacher reminds us, certain emergent ecological imaginaries come to be foregrounded when they perform political and ideological work to produce new registers of value (Rademacher 2011). In and around Gurgaon, the careful machinations of value rely on the entrenched politics of caste and work to secure the allegiance of dominant caste groups who mobilize their authority and power to remap the ecologies of land and water.

In tracking the lives of lost lakes, it then becomes critical to ask what exactly is the urban vision of the future? Is ecological sustainability as a vision even possible in an uneven world? What is at stake in building or expanding cities like New Delhi in arid ecologies of droughts and water shortages and, ironically, floods? What everyday practices of exclusion and dispossession are there, and how do the politics of land and of caste intersect in urbanizing landscapes? These questions are not new, but they have gained renewed urgency and they compel us to attend to a deeply splintered terrain and unravel the contradictions of capitalist urbanization. Answering them prompts us to not only build alliances and rethink the logics of city-making but also find ways to contest ecological loss and create possible avenues to reconcile, as much as possible, urbanization with both ecological sustainability and social justice. Before I turn to discuss the life and death of the Ghata Jheel, I grapple with how lakes and ponds are demarcated and what constitutes a water body.

What Is a Water Body?

In an arid landscape, where water comes and goes seasonally, what qualifies as a water body? What are the attributes of this supposedly watery or not-so-watery object? Must there be water to qualify as a water body? Is a dried-up lake still a lake? How can it be determined, and by whom? These were some of the questions that came up very early on in my fieldwork, as I worked with a GIS student to help me track the contours of Ghata Jheel over the last two decades and develop land classification maps of the lake and its surroundings. It was an interesting exercise and we went back and forth, only to soon realize that mapping land and water is tricky. For instance, even though we had read the reports that the lake had dried up, when we created land classification images using satellite data the area of the same lake was larger in 2014 than it was in 2003. This was puzzling, but the exercise urged us to reflect on our own positivist assumptions and reconsider water bodies not only as stable spatial entities but as social-ecological assemblages that are enmeshed in regional temporalities of seasons, precarities of weather, rainfall, and temperatures. We had to consider the rhythms of land and water and also consider when and how the extensive but tenuous hydrological network that underlies the rocky terrain gets saturated or not.

Mathur and da Cunha, in their design platform *Ocean of Wetness*, have argued that "there is no such thing as dry land" and have drawn attention to the potency of design and to the anthropogenic agency that separates land from water and contours the amphibious, and often indeterminate, wet and dry geographies (https://www.mathurdacunha.com/ocean-of-wetness, accessed April 22, 2020). The local agro-pastoralists, who tended cattle, cultivated crops, grew vegetables, and collected stones to make a living in the rugged landscape of Gurgaon, would concur with Mathur and da Cunha. They were acutely aware of the unreliable nexus of land and water. Even when I asked about land and its transformation, my interlocutors would inevitably turn to talk about the precarity of water, the fluctuations of rain, the direction of flow, the levels of the water table, the slope of land, and the sites of water reservoirs. They would almost always talk about the lake in conjunction with the years of floods and drought—1964, 1975, or 1977, years when they got plenty of rain, followed by a good crop, or when there was a drought and all was parched—highlighting the fragility of the lines that bounded the spaces of land and water. Not surprisingly, there is a vast vocabulary that indexes the degrees of aridity, irrigation, practices of cropping, usufruct, and property. For instance, *chahi* or *abi* is the irrigated land, while *banjar kadim* refers to long fallow, *banjar jadid* refers to short fallow, *banjar mumkin* refers to cultivable water, *gair mumkin* is uncultivable, and *pahadi* is the rocky terrain.

In December 2018, Ragbar accompanied me to the different *johads* (smaller ponds) in the Ghata village and thoughtfully said,

> There are five small johads in the village. All the villagers used to come out and clean the johads and we all maintained them. There was *bhaichara* (brotherhood), we listened to each other and lent a hand but now the municipality has come, so no one takes care of them. There were no borewells [then], the johads made life possible (*mumkin*), our agriculture, soil (*mitti*), and life (*jeevan*) all depended on johad. (Personal communication, December 18, 2018)

The temporal embrace of land and water that makes life possible in arid ecologies prompted us to abandon our purifying attempts to "classify land" or map the boundaries of land and water. While the short foray into mapping land and water was not particularly helpful in the end, it unraveled the boundary work that generates a morphology of water and land, and revealed the need to attend to the vicissitudes of land-water simultaneously and grapple with ecologies in all their social-material complexity. We then turned to geological and meteorological details like the seasonal cycles, amounts of rainfall, length of the monsoon, floods, and droughts, which allowed us to acknowledge both the limit of the map and the potentialities of an arid landscape. Attending to the ecological attributes of land-water meant that I slowly became cognizant of the grammar of cultivability, fertility, life, and livelihood, which in an arid landscape are contingent, and as Ragbar notes, must be maintained through mutual collaboration and cooperation.

The challenge of identifying and demarcating a water body also dogged the Haryana state authorities who set out to identify all water bodies. The Gurugram Metropolitan Development Authority (GMDA), with the support of environmentalist Chetan Agarwal, surveyed all water bodies and in their effort to restore lost water bodies created a definitive inventory of ponds and lakes and assigned unique identification designations (UIDs) to all of them. But, the attempt to record the number of water bodies and assign them UIDs faced a significant challenge. It was not easy to catalog the number of water bodies, and they had to rely on three different sources: the Revenue Record of 1956, the Survey of India Report of 1976, and the High Resolution satellite imagery of 2011–2012. The number of water bodies varied significantly: the first source identified 640 water bodies, the second 519, and the third 647.

Table 7.1: Calculating water bodies

Source of Evidence	Number of Water Bodies
Record of Rights, 1956	640
Survey of India, Tope-sheet 1976	519
High-resolution satellite images (2011–2012)	647
Water bodies that currently exist in the revenue record	251
Water bodies that Gurgaon has lost	389

Source: GDMA 2019.

With such significant variation, the National Green Tribunal[7] noted that only 123 bodies of water were common to all these numbers and that those should be the focus of their restoration project. One of the main participants involved in preparing the inventory of water bodies noted:

> We consulted the revenue reports. Some water bodies may not be in the revenue record but may be mentioned in the Survey of India's topographical map. So, we acknowledged these sources but then we come to the third fuzzy part where a water body is not mentioned in any record, yet when you go on the ground there is water. Then you have to take a call on whether it is a regular water body that got left out or is it just some low-lying area where some water has collected or is it sewage. (Interview with Chetan Agarwal, February 23, 2020)

The Haryana Pond and Waste Water Management Authority defines a pond as "a tank or lake or any other inland water body having an area of 0.5 acres or more, *whether it contains water or not*, and mentioned in revenue records as *talab, johad,*

---

7. The National Green Tribunal was established in 2010 by the government of India to deal with cases relating to environmental protection and conservation.

tank or by any other name and includes green belt and the peripheral catchments areas, main feeder inlet and other inlets, bunds, weirs, sluices, etc. but does not include wet lands as notified by the Government from time to time" (Haryana Pond and Waste Water Management Authority, accessed February 12, 2020, emphasis added). But a water body that is less than 0.5 acres, located in forest land or on private land, is not considered to be a pond. The size of the pond and its "record" in the revenue registers are critical for ascertaining its material existence. There is also a sense of malleability, "whether it contains water or not," that acknowledges the seasonality, rhythm, and temporality that contours the edges of land and water. Yet, ensconced in the positivist hegemony of documentation, tabulation, and evidence, the new goals of pond restoration and associated creation of inventories have tended to overlook the mobility of water bodies, obscured the rhythms of flow, and proceeded, as Ranganathan has argued, to fix the boundaries of land and water (see Ranganathan 2015).

There is a fair bit of uncertainty about how to demarcate the boundaries of water bodies and even what counts as a water body. And this social-ecological confusion proved to be productive in an urbanizing frontier. Amid speculation and property-making, the amphibious submergence areas, lakebeds, seasonal lakes, and wetlands that were crucial for maintaining regional hydrologies came to be overlooked, encroached upon, solidified, and deemed to be land just waiting for development.

In recent years, an impressive body of work on coasts, deltas, and riverscapes has drawn attention to the entangled nature of land-water relations and to the notions of fluidity, sponginess, or permeability that animate such watery terrains (see Amrith 2018; Bhattacharyya 2019; Camargo and Cortesi 2019; Krause 2013; Lahiri-Dutt 2014; Mathur and da Cunha n.d.; Rademacher 2011). While the wet ecologies of riverbeds, bays, and oceans lend themselves to such a relationality, I would, inspired by this scholarship, suggest that arid ecologies of hardy rocks and ridges, too, have to contend with the friction of land and water and take cognizance of their delicate imbrication and seasonal coproduction. The rocky topography of Gurgaon therefore must be situated at the cusp of colonial calculations, irrigation networks, developmentalist dreams, agricultural improvement, and now urban transformation. That is, landscape of land and water are anthropogenically produced and there is nothing natural about disappearing lakes or urban ecologies at large. To materialize the urban vision, the boundaries between land and water have to be stabilized, lakes and ponds dried up, rocks chiselled and flattened, ditches and depressions filled up, and forests cut so that the work of property and its new valuation can take place. This is precisely the boundary work—the act of consolidating, categorizing, and stabilizing—that reifies the distinction between nature and society, land and water, rural and urban, private and commons, legal and illegal, planned and unplanned, and illustrates how ecologies of urbanism are (re)made in the thorny politics of speculation, planning, and property-making, all of which are entwined with regional histories of caste and authority.

Anticipating the City: Assetizing Land and Water

In December 2014, I met RS in his office in the Sushant Tower in Sector 56 of Gurgaon. After working as a travel agent for almost a decade, RS and his friend and now partner set up their real estate business in 2005.

> We did not know much about the land business, initially it was all very confusing. I did not even know about *marla, bhiga*, but my friend is from here and he has contacts in the villages and they helped us a lot. Between 2006 or so and until last year, the market was delirious. There was so much activity, day and night. Rumors were flying about whose land was going to be acquired, or of a new developer who was giving a good price.... When land use is changed from agricultural to commercial, the price of land can easily go up to ten to fifteen-fold or even twenty-fold. In Sector 68, 1 acre of land was sold for Rs. 30–40 lakhs [3–4 million rupees = US$45,000] in 2006, or even less, but in 2012, the price had gone up to six crores rupees [60 million rupees = about US$930,000]. In those days, we could not sleep, villagers were coming in the middle of the night to sell their land, developers and their agents were doing their rounds in the villages. It was just buy and sell, buy and sell. Some people made a lot of money, crores and crores, but not like the big developers like the IREO, L&T, and other big companies, who knows how much they made. (Personal communication, December 15, 2014)

For a few decades now, the story of Gurgaon has indeed been one of "buy and sell," of spectacular gains, lavish wealth, of big and fat Gujjar weddings, and more. Like most urban peripheries, Gurgaon is a frontier of accumulation, charged with anticipation, volatility, hope, but also fear (see Gururani and Dasgupta 2018). Starting in the mid-1980s, it witnessed a dramatic transformation, which I have elsewhere described as agrarian urbanism (see Gururani 2019). With the liberalization of India's economy, urbanization and commercialization of agrarian hinterlands were identified by the government as one of the key strategies to attract global investment and set up business processing offices, special economic zones, highways and expressways, and housing for a growing middle class. In 1985, to make room for the new city, the territorial boundary of National Capital Territory, surrounding Delhi, was extended from 42.7 square kilometers to 1,484 square kilometers to create a National Capital Region (NCR). The NCR, which has since grown considerably, incorporated vast stretches of land in the neighboring states of Haryana and Uttar Pradesh. In Haryana, seven out of nineteen districts, including a third of its villages and towns and about 40 percent of the state's population, are included in the NCR.[8] Due to its proximity to the international airport, Gurgaon became a

---

8. There are now twenty-two districts in Haryana, of which thirteen are in the NCR.

desirable destination and the price of land increased several folds. In this land rush, Haryana opened its doors to private developers, and developers like DLF, Ansals, and Unitech were the first to jump into the land bonanza.[9] With full cognizance and even support of the state, a powerful nexus of local politicians, developers, brokers, real estate companies, and agrarian capitalists worked together to purchase land directly from farmers and put it into the property market. In this conjuncture, the sedimented relations of caste-based authority and power were mobilized, and dominant castes, namely the Jats and Yadavs who are the landowners, brazenly flouted land laws and planning codes, tampered with revenue records, forged land titles, duped smallholders, fought countless court cases, and converted agricultural zones to residential or commercial developments—and turned water into land, and land into gold.

After this initial frenzy, a new round of extended urbanization began in the early 2000s and created what was then called the "New Gurgaon." With the availability of expensive machinery like earthmovers, excavators, loaders, and bulldozers, the plan of making New Gurgaon was conceived on a heavily engineered rocky terrain. Not surprisingly, two key social-spatial technologies—master planning and un-commoning—came in handy once again. Through creative maneuvering of legal, infrastructural, and financial processes, the tools of urban planning not only made agricultural land urbanizable but the plans skillfully manipulated existing regimes of property, especially of the village commons. The commons were gradually fragmented, privatized, and turned into a fungible asset.

Through the 1980s and 1990s, Gurgaon evolved in the absence of a master plan and was largely subsumed into the dictates of the NCR.[10] Between 1981 and 2012, different chief ministers of Haryana granted licenses for developing a total of 8,550.32 acres—but during Bhupinder Singh Hooda's tenure from 2005 to 2012, 20,549.63 acres of land had been granted (Singh 2012). This massive land grab was made possible by the familiar instrument of "master planning." Multiple plans were rapidly crafted, vast stretches of agricultural land rezoned as urban or commercial, revenue records and land titles tampered with, and water bodies erased in an unprecedented attempt to financialize space and create propertied geographies. Typically, city plans and planning commissions have a horizon of ten, twenty, or more years—but not in Gurgaon. Under Mr. Hooda, three plans were drafted between 2005 and 2012.

In July 2006, Gurgaon released its first draft of Master Plan 2021 (DP 2021), which was notified six months later, in February 2007. Under Section 4 of the Land Acquisition Act of 1894, a draft plan is a necessary precursor to a master plan: the draft plan makes public the state's intent to acquire land for public purposes,

---

9. DLF is the acronym for Delhi Land and Finance. With its close connections with the then-ruling Congress party, it acquired thousands of acres of land, changed their land use, consolidated them, developed them, and pursued its dream to *Build India*. See Gururani 2013.
10. The chief minister served as the director of the Haryana Urban Development Authority (HUDA), Town and Country Planning (TCP), and Haryana State Industrial and Infrastructural Development Corporation (HSIIDC) and administered all land transactions.

and it seeks input and acknowledgment. After the draft plan under Section 4 is announced, the land is surveyed under Section 5A, followed by Section 6, which invites objections, and then finally Section 9, which announces the award/compensation for the land that is to be acquired. It is impressive that most landholders I spoke with were fairly conversant with the various sections of the Land Acquisition Act, knew the dates when plans were announced, and kept a keen eye out for Section 4, which would make their land available to the state with little compensation, under eminent domain or not. Given the very high stakes, the period between the draft plan and when the ensuing master plan is notified is a frenzied time of real estate activity. Rumors begin to fly about which zones have been earmarked for acquisition and of how land can be sold to private developers before the final notification. The developers and their many subsidiaries approach local landholders, strike deals, and purchase land that may be announced under Section 4, mostly before the governmental notification, for very low prices but for more than the state compensation. There is a great deal of insider information traded at this moment; as RS noted, "everyone is involved in this (*sabki mili bhagat hoti hai*), from ministers, *patwaris* (revenue officials), brokers, developers, *sarpanch* (village council head), all are involved" (personal communication, December 12, 2015). Draft Plan 2021 was notified within the very short period of six months after it was presented, and as the journalist Shalini Singh noted, "the Master Plan 2021 reflected a dramatic shift from the draft plan, with the land use of as many as 11 sectors changed from public and semi-public/public utility/open space/industrial to residential and commercial. The government did not offer any explanation for the substantial overhaul of this draft plan in the final master plan" (Singh 2013). Before the Master Plan 2021 was released, "there was an over 30-fold surge in the purchase of land after Mr. Hooda took office in May 2005. There was an over Rs. 40,000 crore escalation in the value of roughly 1,000 acres of land through MP 2021 alone" (Singh 2013). With hidden kickbacks built in and soaring land prices, land became incredibly valuable. Only four years later, in October 2010, another draft of the Master Plan (now 2025) was released and was notified promptly in May 2011, and between the two Master Plans (MP2021 and MP2025), the state government further converted roughly 500 acres of agricultural land and forests into residential land use. Soon after, in November 2012, just over a year later, a third draft, for Master Plan 2031, was released and was notified in September 2013. In the Master Plan 2031, sixty new sectors were added and, amid intense speculation and anticipation, the Department of Town and Country (DTCP) classified the hitherto agriculture lands as urbanizable and issued several licenses to developers to urbanize land.

In tracing the history of city-making through the *chars* (marshes) of colonial Calcutta, Bhattacharyya in her book, *Empire and Ecology in the Bengal Delta*, draws attention to the mobility of a swampy landscape, of its multiple indeterminacies, and demonstrates how property was crafted in the crucible of colonial power. She maps the legal processes and situated ecological temporalities of a deltaic region and writes that, "urban marshes, floating watery soils and riverine sedimentations were not always property, but had to be made into property," a complex legal and technological venture that preoccupied colonial speculators, developers, lawyers, and engineers (Bhattacharya 2019, 4). Bhattacharyya considers property as a process, "a form to frame our thinking about land, as a language to express the division between land and water and as a legal technology to demarcate land, marsh, accretion and water" (Bhattacharya 2019, 23). In drawing on Bhattacharya in a very different topographical context, I would suggest that ponds, ditches, and lakes, too, came to be seen as "land-in-waiting for property development" (Bhattacharya 2019, 169). Instead of seeing them as "impediments" that thwarted suburban expansion, in Master Plan 2021 the lakebed of Ghata Jheel was rezoned and designated an urbanizable residential sector, number 58. Several developers like BPTP, Vatika, Ansals, DLF, Emaar-MGF, IREO, and their corporate subsidiaries purchased licenses: 651 acres of land were licensed to IREO, half of which was Ghata Jheel's lakebed. In this transaction, from being a relatively small player, IREO almost overnight emerged as a recognizable brand name in India and has since built a state-of-the-art gated residential enclave called The Grand Arch.

CS, a middle-aged resident of Ghata village, who had earlier served as a member of the village panchayat said:

> Huge trucks and bulldozers started to come. They worked day and night, breaking the *pahadi*, crushing rocks, filling, and raising the Jheel. They brought their men, it all happened in front of our eyes. . . . Earlier there was no value (*keemat*) of land here. No one ever offered us money (for land). The soil is not good here, we can barely grow one crop, and there was always a threat of flooding. When companies started to come, everyone wanted to sell, everybody wanted money. We hardly negotiated, some may have made good money, but most of us went in "*ghata*" (loss). The lakebed was *shamalat* (commons), it was commonly used by all and managed by the *panchayat* (village council) but some *kashtkar* families claimed their share and sold it to the Company. Now there is a court case going on. (Personal communication, December 15, 2018)

In an infrastructural feat involving heavy machinery, the vast submergence area of the Jheel that had stretched over a kilometer and covered roughly three hundred acres was re-engineered; covered up, erased, or flattened; licensed for development; and turned into solid property/land for the city to rise.

While the technologies of planning reworked the boundaries of land and water and laid the foundation for a new urban space and imaginary, it was the legal registers of property, especially of the commons, that had to be rescripted. Property rights in colonial Punjab, of which Haryana was a part, have a long and convoluted history that, as the historian Neeladri Bhattacharya has argued, were hard to categorize (Bhattacharya 2019, chapter 4). Yet, over time, caught within colonial regimes of revenue, tenure, shares, tenancy, and customary practice, a system of property rights and obligation was created that vested authority in the village proprietary body (*malikan deh*), which was primarily constituted of the dominant caste of Jats and, in some regions, of Yadavs.

Space does not permit a comprehensive illustration here, but Bhattacharya has argued that the complex reconfiguration of three main tenurial arrangements during the British rule—*zamindari*, *pattidari*, and *bhaichara*—were central to stabilizing the contours of a village society. In particular, it was the invented notion, according to Bhattacharya, "of a bhaichara tenure as one in which rights were regulated by collective custom, as distinct from pattidari, a tenure in which ancestral shares defined rights and obligations" was critical (Bhattacharya 2019, 117). It was an intriguing contortion that consolidated the authority of the powerful landowners, as it "made . . . control over a holding that was larger than their ancestral share legitimate and legal" (Bhattacharya 2019, 121). At the same time, as Chakravarty-Kaul (1996) notes, colonial revenue settlements created property rights that vested authority in the village proprietary body (*malikan deh*) and in the colonial state and gave the village proprietary body (formed from dominant castes) the authority to collectively hold rights over the cultivable fallow, residential areas, and the *johads*. But, as CS pointed out, even though the proprietary rights were consolidated in favor of the dominant castes, all villagers maintained their user rights in the commons for pasture, fodder, and cattle. Tensions prevailed between the proprietors and the non-proprietors but by and large in Ghata, until recently, the lakebed was used for common purposes.

Against this messy backdrop of property rights, two important legislations were passed after independence—the Land Reform Act of 1954 and the Punjab Village Common Land Act of 1961—which made important legal interventions as they vested the authority of the commons in the *panchayat* (council), and took away from the proprietors the right to partition land and privatize the commons (see Kumar, Bren, and Ferguson 2000). In the context of the urban land rush, however, through creative interpretation of competing legal regimes, parts of the Ghata lakebed were designated as "private," and landholders, mostly from dominant castes, partitioned and sold them to private developers. As one local environmentalist pointed out, "once it is deemed private, it is removed from the governmental register of revenue. If it is not in the revenue register, then it is also not on the map, and if there is no documentary record, then a water body ceases to exist. It is gone, disappeared!" (personal communication, January 2020).[11] In the crucible of master planning, speculation, complicated colonial legacy of property regimes, village and caste politics, the watery body of the Ghata Jheel came to be slowly but surely erased, flattened, hardened, partitioned and turned into land, and assetized to usher in the city of tomorrow.

---

11. There is currently a court case going on between the two proprietors and the village proprietary body, Punjab and Haryana High Court Case No. 6590, Gram Panchayat Ghata v. State of Haryana, related to ownership of common lands.

Figure 7.1: Squatting and recycling in Ghata Lake–bed by migrant workers

Speculating Sustainability: Lush Landscape of Water and Steel

Ghata Lake is not the only lake that has disappeared. As mentioned above, over one hundred lakes have gone missing in Haryana alone, and a similar story of disappearing lakes seems to be repeated in the peripheries of most metropolitan cities in India, such as Mumbai, Bengaluru, and Chennai. In response to growing concern over the environmental threats posed by diminishing or disappearing water bodies, the National Green Tribunal (NGT) directed governmental authorities to monitor the state of water bodies and file a report. In addition, residents and environmental groups have been galvanized to challenge the environmental harm done to local forests and wetlands. In Gurgaon, in 2018, Lt. Col. S. S. Oberoi filed an application before the NGT, demanding that the Ghata Lake be restored and no further permission be granted "to develop over waterbodies and their catchment area. It also sought direction to restore 214 other waterbodies and natural channels in Gurgaon District and similar waterbodies in Faridabad District" (NGT).[12] This led NGT to pass the Haryana Pond and Waste Water Management Authority Act and create the Haryana Pond and Waste Water Management Authority (HPWWMA, http://hpwwma.org.

---

12. Original Application No. 325/2015: Lt. Col. Sarvadaman Singh Oberoi v. Union of India & Others National Green Tribunal Order, dated July 20, 2018.

in/index.aspx). The Pond Authority, through a painstaking exercise as mentioned above, compiled data over the last several decades and prepared an impressive *Report on Waterbodies of Gurugram* (Gurugram Metropolitan Development Authority 2019). While the steps to document and record the boundaries of water and land must be applauded, the task of enumerating also raises questions about the challenges of counting and stabilizing the dynamic ecological processes and about the possible erasures and oversights such an exercise may entail. This, however, is beyond the scope of this discussion.

With the setting up of the Pond Authority, there seems to have been a commitment to restore and regenerate water bodies, but it is caught up between different administrative departments—the forest department, the irrigation department, the urban development authority—with overlapping and sometimes conflicting authority over the same piece of highly valuable land. The boundaries of a water body remain in question and thus far no steps have been taken to revive or restore the Ghata Lake. Instead, the Pond Authority has proudly embarked on an ambitious project to develop eighteen model ponds with the aim of beautification, conservation, irrigation, and compensation for the lost nature. Claiming to be a pioneer in urban sustainability, Haryana is now set to carve one model lake in each of its districts (Bhatia 2019). And, in place of the Ghata Jheel, according to the Regional Officer of the Haryana State Pollution Control Board (HSPCB), "we are . . . building a lake near Sector 72 that will be a large lake, it will be a modern lake that will work like other natural lakes and help address Gurgaon's water problem" (personal communication, December 17, 2018). Relying on a standard repertoire of images and a global discourse of greening, sustainability, adaptation, and ecological futurity, the Pond Authority highlights beautification and restoration of natural heritage as one of its key goals in manufacturing new ponds. While the Pond Authority is willing to acknowledge ecological loss and disappearing lakes, it is reluctant to join hands with urban development authorities and to ecologically rethink the future of new cities. Ensconced in the binaries of society-nature, rural-urban, and land-water, the Pond Authority has resorted to techno-managerial fixes and to making new bodies of water as a matter of fact; as Bruno Latour would put it, unmoored from its social entanglements and oblivious to the concerns of smallholders who depend on ecologies of land and water for their lives and livelihoods (Latour 2005).

It is not only the state that has embraced the discourse of sustainability, harmony, and greenness; the developers, too, have responded to the green call. The real estate developer IREO, which incorporated parts of the Ghata Jheel and built a state-of-the-art luxury residential enclave and is currently developing an IREO city, offers "lush landscapes" and "acres of carefully designed landscape [that] will complement the glorious natural surroundings that encircle The Grand Arch" (IREO n.d., https://www.ireonewprojects.net/pdf/1518-Grand%20Arch%20Brochure%20 3%20parts_Brochure%2001-09.pdf).

In IREO's ode to ecological concerns, the centerpiece of the Grand Arch is a twenty-two-foot-high metal tree, called the "Dada" tree, that has been designed by the famous artist Subodh Gupta.[13] The tree resembles a banyan tree and is constructed out of polished, hyper-reflective steel, with curvy branches, and has clusters of utensils as leaves. Leaving aside the fact that a tree is a brash embodiment of patriarchy, male inheritance (*dada lai*), and masculinity, what such artifacts of nature accomplish, materially and discursively, is to bolster the developmentalist agenda in the cities of the south, generate consensus, and conceal the contested terrain of land and water on which they stand. The tree of metal and the model ponds not only script a new ecological imaginary and stage a new grammar of urban ecologies—albeit ungrounded and superficial—but they also symbolically and materially undermine and erase rurality and agrarian histories of the waterland that surrounds them.

In invoking the discourses of ecological restoration and natural heritage, the project of city-making boldly and unabashedly erases sedimented socio-ecologies and their agrarian entanglements, temporalizes space, overlooks the fragile embrace

Figure 7.2: View of the Grand Arch from Ghata Village. Photo by author, December 2018.

---

13. According to Gupta, the Dada tree symbolizes Indian heritage and familial ties, as connoted by the kin term—*Dada*, which in Hindi means grandfather. But in another iteration, it also means a bully and also expresses patriarchal assertion and authority (Pandey 2015).

Figure 7.3: Facing Grand Arch. Photo by author, December 2018.

of land and water, and works to commodify nature, assetize land, and privatize property. Yet, the force of flood, as seen in July 2016, belies such valiant attempts and urges us to remember that, as KS said, "in place of a village, they can build buildings and make a city but a flood does not distinguish, you can try and stop [water's] course (but) we cannot ignore its fury. This is the nature's game, this is life, and it is up to us to acknowledge it or get washed away" (personal communication, December 2018).

Conclusion

Taking KS's comments seriously and not ignoring nature's game, I have focused on lost lakes to show how emerging urban geographies of value are crafted through socio-spatial technologies of planning and de-commoning that demand a firm ground, a stable terrain that can materialize the dreams and desires of speculation and turn water into landed (private) property (see Bhattacharyya 2019, 23–31). I have drawn attention to the imbrications of land and water in an arid terrain and described how the uneven sedimentations of property and caste come to inscribe the registers of capitalist urbanization, shape new ecological imaginaries, and craft new urban futures. And, as water is turned into land, water seasonally asserts its presence through newly destructive channels. Flooding in Gurgaon, which is becoming a regular feature of the monsoon, is a reminder that if, we wish to move "beyond the mess created by an anthropocentric hubris that informed the conquest of nature in modern times," as the editors of the volume urge us to, then it is critical that we rethink our disciplinary silos that separate urban from rural, water from land, and ecologies from society (Rademacher and Sivaramakrishnan, this volume)[14] and attend carefully to the grounded rhythms of socio-ecological relationships and embedded ecologies that can make or break the future of urbanizing frontiers in Asia.

Acknowledgments

I am grateful to the participants in the two workshops, "Death and Life of Nature in Asian Cities, I and II," held at the Hong Kong Institute of Humanities and Social Sciences and Yale University respectively, and especially to Anne Rademacher and K. Sivaramakrishnan for their invaluable comments on earlier versions of the chapter. My special thanks to Amy Zang and Kasia Paprocki for their insightful commentaries that helped improve the paper immensely. I would also like to thank all the interlocutors in Gurgaon who shared their stories with me, to Rituparna Sengupta for her research assistance, and last but not least to Rabab and Harsh Lohit for hosting me and for making fieldwork possible in multiple ways.

Works Cited

Agnihotri, Indu. 1996. "Ecology, Land Use and Colonisation: The Canal Colonies of Punjab." *The Indian Economic & Social History Review* 33 (1): 37–59.

Amrith, Sunil. 2018. *Unruly Waters: How Mountain Rivers and Monsoons Have Shaped South Asia's History*. New York: Basic Books.

Arora, Shilpy. 2018. "Gurugram Lost 389 Water Bodies in 60 Years: Study." *The Times of India*. January 26, 2018. https://timesofindia.indiatimes.com/city/gurgaon/gurgaon-lost-389-water-bodies-in-60yrs-study/articleshow/62610956.cms.

Bear, Laura, Ritu Birla, and Stine Simonsen Puri. 2015. "Speculation: Futures and Capitalism in India." *Comparative Studies of South Asia, Africa and the Middle East* 35 (3): 387–91.

Bhatia, Varinder. 2019. "Explained: Haryana's Johads Set for Revamp, 18 Model Ponds Planned." *The Indian Express*. November 25, 2019. https://indianexpress.com/article/explained/explained-haryanas-johads-set-for-revamp-18-model-ponds-planned-6135128/.

---

14. See Anne Rademacher and K. Sivaramakrishnan, "Introduction: Urban Nature Brought to Life in an Age of Loss," in this volume.

Bhattacharyya, Debjani. 2018. *Empire and Ecology in the Bengal Delta: The Making of Calcutta*. Cambridge: Cambridge University Press.

Bhattacharya, Neeladri. 2019. *The Great Agrarian Conquest: The Colonial Reshaping of a Rural World*. Ranikhet, India: Permanent Black.

Camargo, Alejandro, and Luisa Cortesi. 2019. "Flooding Water and Society." *Wiley Interdisciplinary Reviews: Water* 6 (5): 1–9.

Chakravarty-Kaul, Minoti. 1996. *Common Land and Customary Law: Institutional Change in North India over the Past Two Centuries*. Delhi: Oxford University Press.

Choudhry, Chetna. 2016. "After 20-Hour Gurujam Nightmare, Now Haryana-Delhi Blame Game." *The Times of India*, July 30, 2016. https://timesofindia.indiatimes.com/india/After-20-hour-Gurujam-nightmare-now-Haryana-Delhi-blame-game/articleshow/53458206.cms.

Chowdhury, Prem. 1986. "The Advantages of Backwardness: Colonial Policy and Agriculture in Haryana." *The Economic and Social History Review* 23 (3): 263–88.

Das, Veena. 1995. *Critical Events: An Anthropological Perspective on Contemporary India*. New Delhi: Oxford University Press.

Goldman, Michael. 2011. "Speculative Urbanism and the Making of the Next World City." *International Journal of Urban and Regional Research* 35 (3): 555–81.

Government of India. 2010. *National Disaster Management Guidelines: Management of Urban Flooding*. National Disaster Management Authority. ISBN: 978-93-80440-09-5. New Delhi.

Government of Punjab. 1883–1884. *Gazetteer of the Gurgaon District. 1883–84*. Compiled and published under the Authority of Punjab Government. Lahore: Arya Press.

Gurugram Metropolitan Development Authority (GDMA). 2019. *Water Bodies of Gurugram: A Report*. Accessed February 12, 2020. http://www.indiaenvironmentportal.org.in/content/463595/report-on-waterbodies-of-gurugram-district/.

Gururani, Shubhra. 2013. "Flexible Planning: The Making of India's 'Millennium City,' Gurgaon." In *Ecologies of Urbanism in India: Metropolitan Civility and Sustainability*, edited by Anne Rademacher and K. Sivaramakrishnan, 119–43. Hong Kong: Hong Kong University Press.

Gururani, Shubhra. 2019. "Cities in a World of Villages: Agrarian Urbanism and the Making of India's Urbanizing Frontiers." *Urban Geography*. October 22, 2019. doi: 10.1080/02723638.2019.1670569.

Gururani, Shubhra, and Rajarshi Dasgupta. 2018. "Frontier Urbanism: Urbanisation beyond Cities in South Asia." *Economic and Political Weekly* 53 (12): 41–45.

Harms, Erik. 2012. "Beauty as Control in the New Saigon: Eviction, New Urban Zones, and Atomized Dissent in a Southeast Asian city." *American Ethnologist* 39 (4): 735–50.

Haryana Pond and Waste Water Management Authority. Accessed February 12, 2020. http://hpwwma.org.in.

IREO. n.d. "The Grand Arch: Luxury of Location." Accessed May 2018. www.ireonewprojects.net/pdf/1518-Grand%20Arch%20Brochure%203%20parts_Brochure%2001-09.pdf.

Krause, Franz. 2013. "Seasons as Rhythms on the Kemi River in Finnish Lapland." *Ethnos* 78 (1): 23–46.

Kumar, Ajai, Leon Bren, and Ian Ferguson. 2000. "The Use and Management of Common Lands of the Aravalli, India." *International Forestry Review* 2 (2): 97–104.

Kumar, Ashok. 2012 (2017). "137 Water Bodies Have Dried Up in Gurgaon." *The Hindu*, October 15, 2012. Updated June 5, 2017. http://www.thehindu.com/todays-paper/tp-national/tp-newdelhi/137-water-bodies-have-dried-up-in-gurgaon/article12558240.ece.

Lahiri-Dutt, Kuntala. 2014. "Beyond the Water-Land Binary in Geography: Water/Lands of Bengal Re-visioning Hybridity." *ACME: An International E-Journal for Critical Geographies* 13 (3): 505–29.

Latour, Bruno. 2005. *Reassembling the Social: An Introduction to Actor-Network Theory*. New York: Oxford University Press.

Mathur, Anuradha, and Dilip da Cunha. *Ocean of Wetness: A Platform for Design*. Accessed April 22, 2020. https://www.mathurdacunha.com/ocean-of-wetness.

Mozingo, Louise A. 2011. *Pastoral Capitalism: A History of Suburban Corporate Landscapes*. Cambridge, MA: MIT Press.

Narain, Vishal, and Aditya K. Singh. 2017. "Flowing against the Current: The Socio-Technical Mediation of Water (in) Security in Periurban Gurgaon." *Geoforum* 81 (May 1): 66–75.

Pandey, Parul. 2015. "22-Foot-High Steel Tree Installed in Gurgaon." *The Times of India*, April 28, 2015. https://timesofindia.indiatimes.com/entertainment/events/gurgaon/22-foot-high-steel-tree-installed-in-Gurgaon/articleshow/47072036.cms.

Rademacher, Anne. 2011. *Reigning the River: Urban Ecologies and Political Transformation in Kathmandu*. Durham, NC: Duke University Press.

Ranganathan, Malini. 2015. "Storm Drains as Assemblages: The Political Ecology of Flood Risk in Post-colonial Bangalore. *Antipode* 47 (5): 1300–20.

Roychowdhury, Anumita, and Shubhra Puri. 2017. "Gurugram: A Framework for Sustainable Development." New Delhi: Centre for Science and Environment.

Sengupta, Sushmita. 2016. "Why Urban India Floods: Indian Cities Grow at the Cost of Their Wetlands." *Down to Earth Publication*. New Delhi: Center for Science and Environment.

Singh, Shalini. 2012. "Behind Haryana Land Boom, the Midas Touch of Hooda." *The Hindu*, October 30, 2012. Updated October 31, 2012. http://www.thehindu.com/news/national/behind-haryana-land-boom-the-midas-touch-of-hooda/article4048394.ece.

Singh, Shalini. 2013. "Builder Profits Soar as Master Plans Proliferate in Gurgaon." *The Hindu*, May 27, 2013. https://www.thehindu.com/news/national/builder-profits-soar-as-master-plans-proliferate-in-gurgaon/article4753735.ece.

Sohail-Hashmi. 2016. "Why the Gurgaon Deluge Is Only a Taste of Things to Come." *HuffPost*. https://www.huffingtonpost.in/2016/07/31/why-the-gurgaon-deluge-is-only-a-taste-of-things-to-come_a_21442097/.

Thorner, Daniel, ed. 1996. *Ecological and Agrarian Regions of South Asia Circa 1930*. Karachi: Oxford University Press.

Williams, Raymond. 1973. *The Country and the City*. Oxford: Oxford University Press.

8
Concrete Ecology: Covering and Discovering Saigon's Ecology in a Time of Floods

Erik Harms

A Flood of Discourse

In late 2010 and early 2011, while conducting ethnographic research on the social life of master-planned developments in Ho Chi Minh City, I noticed that residents across the city were regularly discussing the city's growing problem with urban flooding, which was getting noticeably worse with each passing year.[1] Commentary on flooding certainly isn't new to Vietnam. In precolonial times, imperial annals occasionally referenced floods or droughts when assessing an emperor's legitimacy and benevolence, or lack thereof (Kiernan 2017, 148–50). Later, French colonial officials periodically justified their civilizing mission by disparaging Vietnamese water management (cf. de Grammont 1863, 93–95). Still later, as if giving the French a dose of their own civilizing medicine, future leaders of the Vietnamese Communist Party forcefully denounced incompetent French flood management practices in the 1938 anticolonial tract, *The Peasant Question* (Truong Chinh, and Vo Nguyen Giap [1937–1938] 1974, 85–89). Invoking floods to complain about the powers that be—or to claim that those in power *shouldn't* be—is nothing new.

Throughout the second decade of the new millennium, a tide of politically inflected flood talk again started rising across Saigon. Internet meme-makers circulated doctored images of the flooding that mocked government slogans, and everyday residents across the city expressed their own anger by criticizing urban and national government. This chapter traces some of the currents in this flow of discourse to show how flood talk is not only a common part of everyday conversations in contemporary Saigon but prompts people from all walks of life to engage in debates about urban development, especially about transformations to the built landscape. When people complain about floods, they are not only mad at the water but mobilize flood talk to express their disagreement about the ways other human

---

1. These conversations echoed scientific research appearing around the same time. For example, Nicholls et al. (2008) ranked Ho Chi Minh City among the world's ten cities most vulnerable to climate change, and a widely cited 2007 World Bank report identified Vietnam as one of the five countries in the world most vulnerable to the impacts of sea level rise (Dasgupta et al. 2007, 40).

beings are choosing to construct the city they live in. Commentary on unruly water flows directly into politicized debates about what (and who) is causing the flooding, and how people should go about building and living life in a city increasingly beset by disruptive inundations. In this way, as Danny Marks has argued in his work on the 2011 floods in Bangkok, urban flood events must be understood as distinctly socionatural phenomena, which not only inundate a city but become entangled in politics and transform the very way in which people understand the meaning of urban nature and life (Marks 2015). Like floodwaters themselves, the currents of knowledge and debate don't just flow randomly but instead interact with the material world, swirling around and forming eddies and patterns in the channels created by the intersection of the natural and built landscapes. This chapter follows a selection of these socionatural currents of water and discourse, focusing specifically on how flood talk and other forms of environmental discourse stream into and then intermingle with debates about the role of master-planned mixed-use development projects called New Urban Zones. Depending on who one listens to, these urban development projects can be understood either as a major *cause* of increased urban flooding or as part of the *solution*. For everyone, however, New Urban Zones play an outsized role in shaping the ways people think about urban nature and ecology in the city.

Three Currents of Debate

Over the past decade, Vietnamese newspapers have tended to blame urban flooding on climate change and rising sea levels. However, most Saigon residents point the finger primarily at New Urban Zones, which have been (and still are) filling in the canals and rivulets of low-lying lands on the outskirts of the city. The logic informing the critiques is quite straightforward. Building New Urban Zones at the scale in which they are being built in Saigon requires moving mounds of earth and pumping sand and infill into wetlands. Everyone in the city can see the sand barges lined up along the Saigon River, constantly pumping slurry sand into the wetlands (Figure 8.1). To my friends in the city, the causal connection between the city's rapid urban expansion and flooding is obvious; as new developments continue to rise out of low-lying peri-urban swamps, so too do the flood waters in the rest of Saigon. While not based on formal scientific research, the causal connection people make operates as a "science of the concrete," a concept Claude Lévi-Strauss developed in order to show how vernacular reason often shares an affinity with scientific reasoning (Lévi-Strauss 1966). Just like scientists, city residents readily identify a clearly observable cause (filling in wetlands with urban developments) and link it to a clearly observable effect (flooding).

Meanwhile, the designers and planners of New Urban Zones speak in a highly technocratic language that purports to mobilize fact-based science and environmental reason but ignores the concrete causal logic described by everyday residents.

Figure 8.1: Pumping sand into wetlands in order to build the Thủ Thiêm New Urban Zone. Source: author, November 2010.

The "science talk" urban designers use to describe their development projects is conveyed abstractly, referencing technocratic terminology like "sustainable urban development challenges," "integrated sustainability strategies," "heat gain," "ventilation," "greenroofs," "stormwater runoff," "air movement," "sedimentation," "soil geography," and "concrete hardsurface." Unlike most city residents, however, urban designers mobilize these terms to justify rather than to critique the New Urban Zones they have been charged with designing. In the process, this kind of discourse produces an abstract concept of "urban ecology" that borrows from the vocabulary of eco-consciousness but fundamentally ignores the concrete realities of Saigon's ecosystem. The conception of ecological design and sustainability that emerges is one that sounds like or evokes the idea of "ecology" to justify the terraforming of land and the eviction of large numbers of people. This abstract understanding of ecology is decidedly different from the concrete understanding of ecology espoused by everyday residents.

As I will show in this chapter, although the language of planners deploys "scientific" sounding rationalization, the logic of the plan they have proposed departs dramatically from the conclusions actual urban scientists have reached in their work on urban ecology. The scientists, it turns out, hold an understanding of the situation much more aligned with everyday city residents than with the ideas promoted by professional planners and designers. As evidence for this argument, this chapter focuses on three discernable currents of debate associated with three different sets of actors who simultaneously contribute to and become caught up in the flood of discourse about New Urban Zones. First, I discuss the conclusions of self-proclaimed practitioners of hypothesis- and data-driven science who publish in peer-reviewed journals. Second, I discuss the views of everyday citizens who do not call themselves "scientists" but nevertheless engage in the science of the concrete. Third, I show how these perspectives relate not only to each other but to those of urban planners who use abstract technological language to lend their assertions a scientific credibility while ignoring the obvious ecological effects of their plans. Like the intermingled waters of a flood, these discursive currents may blur into each other at the edges, but they can nonetheless be understood as unique streams of thought animated by social actors with identifiably different ways of thinking about urban development. The rest of this chapter traces these three currents of debate, showing how the citizens and the scientists have ultimately come to the same conclusion. Meanwhile, the urban designers differ dramatically, as they continue to try and mobilize abstract scientific concepts to promote an agenda that contravenes the very logic of urban ecology they claim to be working to protect.

What the Scientists Say

What is causing the flooding in Saigon? While some blame must of course fall on the broad problem of climate-induced sea level rise (SLR), in paper after peer-reviewed paper, scientific researchers have demonstrated that the bulk of flooding stems from the immediate effects of urbanization. For urban contexts, scientists specifically highlight the convergence of two factors. The first factor is *subsidence*, caused primarily by the combined effect of draining the water table and overbuilding. The second factor is the *expansion of impermeable surfaces*, due to extensive paving with macadam (asphalt) and concrete. In colloquial terms, the convergence of these two factors means that Saigon is slowly sinking because a thirsty and ever-expanding urban population is both emptying aquifers and constructing heavy things like roads and buildings that weigh down upon the very ground that has been hollowed out. To make matters worse, the extensive use of pavement and concrete surfaces has made it increasingly difficult for excess waters to seep back into the soil and recharge the aquifers. Left hollowed out, the ground upon which the city is built succumbs to further subsidence. With nowhere to go, the surface water turns into rivers of runoff that rage right through the city itself (Tran Thi Van, Bui Thi Thy Y, and Ha Duong Xuan Bao 2015). In a typical statement on the matter, one scientific assessment of flood risk in Ho Chi Minh City quite clearly explains that the "influence of planned urban developments to the year 2025 on future flood risk is seen to be significantly greater than that of projected sea-level rise to the year 2100" (Storch and Downes 2011, 517). In another paper, the conclusions are similar: flooding in the city provides "evidence of inappropriate urban expansion leading to increase in flooding vulnerability. Although climate change impacts are obvious, the rapid population growth and associated accommodation [of] development are believed to be the key cause which has not been solved" (Duy et al. 2018, 209).

Scientific researchers concur that sea levels are certainly rising due to climate change, but not fast enough to account for the problem of urban flooding seen in cities around the world. When it comes to floods, the real threat to the city is neither "nature" nor "climate" in some abstract disembodied sense; instead, the threat results from anthropogenic transformations of nature, which in turn transform what it means to be a human being living in the city. As Rademacher and Sivaramakrishnan have shown in the introduction to this book[2] and in previous research, any study of urban ecologies must not only engage with the way "nature" exists in the city or is threatened by it but how cities themselves both emerge from and transform their own urban ecologies (Rademacher 2015; Rademacher and Sivaramakrishnan 2013). In the case of urban flooding, the city's own growth—both made possible and constrained by everything from politics to economic pressures to social processes to ecological affordances and limitations—has become its own worst enemy (Garschagen 2015, 606).

None of this is unique to Saigon, but the problem is especially acute in Asia. In this volume alone, chapters on flood-related deaths in Mumbai (Solomon, chapter 5),[3] and the interaction of cities like Kolkata and Dhaka with flood landscapes of the Sundarbans (Paprocki, chapter 3)[4] all look to flood landscapes as a way of tracking the socionatural confluence of ecological change, environmental discourse, and social life. Such a focus is not surprising—clearly the literature is responding to very real biophysical processes transforming social and ecological life in these cities. Scholars are increasingly writing about flooding because flooding is increasingly a problem, but flooding is primarily understood as a "problem" to us because of the way it impacts our social and material existence. As Michelle Miller and Mike Douglass note, ecological location alone cannot explain the problem of flooding: "Floods in the rapidly urbanizing settings of Asia are typically triggered by a complex combination of seemingly natural sources (weather events, tsunamis, earthquakes, landslides, and coastal erosion) that are increasingly inseparable

---

2. See Anne Rademacher and K. Sivaramakrishnan, "Introduction: Urban Nature Brought to Life in an Age of Loss," in this volume.
3. See Harris Solomon, "The Absent Presence: Potholes in Urban India," chapter 5 in this volume.
4. See Kasia Paprocki, "The Village at the End of the World: Ecologies of Urbanism in Climate Crisis Imaginaries," chapter 3 in this volume.

from anthropogenic factors" (Miller and Douglass 2015, 505). Similarly, Matthias Garschagen's work on flooding in the Mekong Delta city of Cần Thơ shows that the *combined* interaction of humans and nature form the essential components of urban flooding (Garschagen 2015, 600).

Humans obviously did not invent Southeast Asia's low-lying deltas, but they did choose to settle in them, often because of the affordances provided by the deltas themselves—such as navigable waterways, rich alluvial soils, and abundant food sources. Nevertheless, in the process of capitalizing on those affordances, human settlements have proceeded to transform those very deltas in ways that go on to change the human experience of living in them, in some cases transforming affordances into nuisances or problems to be overcome. The problem is especially acute in coastal and riverfront cities across Southeast Asia, where the interaction of spatial relations with ecological factors (like proximity to the sea, rivers regulated by diurnal tides, and the role of mangroves, swamps, and marshlands) both impacts and is transformed by anthropogenic factors. A recent article comparing Ho Chi Minh City, Bangkok, Manila, and Jakarta reiterates the scientific conclusion that human-induced urban infilling and subsidence are much more likely contributors to the current rise in urban flooding than the more commonly blamed problem of climate change and rising sea levels (Yarina 2018).

Across Asia, environmental concerns are used to justify socioeconomic forms of exclusion (Harms 2016b, 51–52). In addition to the other chapters in this volume, Anne Rademacher's work on Kathmandu described the management of riverbanks in general terms as forms of "socioenvironmental inclusion and exclusion" (Rademacher 2011, 57). Writing of water management in Mumbai, Nikhil Anand and Lisa Björkman have both shown how the expansion of the urban landscape into former marshes and wetlands is precisely a form of socioenvironmental interaction that leads both to political contestations over citizenship and environmental transformation. Narrowing and fixing streams and rivers has also intensified the problem of urban runoff, while also impeding the delivery of potable water (Anand 2017, 129–30; Björkman 2015, 110). The problem of water management cannot be attributed to any preexisting "natural" lack or excess of water but is instead a dynamic interaction between the human and the nonhuman qualities of the ecosystem. Anand notes that "water and its situated infrastructures, as assemblies of the human and the nonhuman, have political effects" (Anand 2017, 230). Furthermore, as Jerome Whitington has noted based on work in Bangkok, the rise of modernist interventions designed to tame the flow of water—like massive concrete walls—"also fractures socio-natural relations" (Whitington 2019, 8).

Everyday Understandings of Flood Ecology

Few Saigon residents would be surprised by any of the conclusions described above. Most residents have a clear conceptual understanding of the basic principles of the city's topography and hydrology, and they commonly described the city as being organized in terms of "low land" (đất thấp) and "elevated land" (đất cao). For example, on one of my first extended research visits to Saigon, in the summer of 2000, a new acquaintance took me on a long motorbike ride through the city. As he showed me around, he pointed out the high and low areas of the landscape, told me which areas were prone to flooding, which areas never flooded, and then proceeded to construct from these differences a social map of urban wealth and poverty that roughly corresponded to vulnerability to rising waters. That same summer, on a car trip around the city's peri-urban fringes, a professor at the University of Social Sciences and Humanities took extra care to tell me about the importance played by the relative land elevation of different parts of the city: poor folk lived in low land and the most well off lived in elevated areas.

These spatial conceptions are not only sociologically meaningful but correspond quite clearly to topographical and soil maps of the city depicted in formal scientific literature (e.g., Dinh Ho Tong Minh, Le Van Trung, and Thuy Le Toan 2015). In the city's low-lying lands, located largely to the city's south and southwest, and also to the east across the Saigon River, the water is called "nước phèn," meaning that it is high in alum content. In other low-lying lands to the southeast, in the direction of the East Sea (South China Sea), mangroves thrive but humans do not because the water is understood as brackish (nước lợ). The implicit geographical knowledge people have of the metropolitan region corresponds to actual topography and maps onto socioeconomic class: the low-lying areas have long been understood as relatively undesirable spots for human life—not only inconducive to drinking but also bad for agriculture.

These understandings of the city's ecology are also linked to stereotypes about the ascribed moral qualities of people who live in these different spaces. Historically, the low-lying lands—especially the mangroves in Cần Giờ district and the swamps in Nhà Bè district—were known as spaces of escape for urban bandits, revolutionaries, and other people living on the margins of state power.[5] At the same time, these areas have always been recognized as natural escape routes for rainwater and the tidal surges of the Saigon River—precisely because of the qualities that prevented them from becoming heavily populated. Ecologically, they are "green lungs" (lá phổi xanh) to the city. Socially, they are understood as underdeveloped, morally suspicious, or both (Harms 2011). In recent years, with some of these peri-urban lands being filled in by New Urban Zones and roads, the developments are not only changing the ecology of the city but its symbolic geography as well. Former "edges" inhabited by marginalized populations are now becoming home to elite and rising middle-class households (Harms 2016a). A kind of double displacement ensues:

---

5. For example, the notorious Bình Xuyên, a group commonly referred to in the Vietnam War literature as a "mafia organization," but who for a time in the 1940s and into the 1950s actually staked serious claims to sovereignty in the south, operated throughout Saigon-Cholon but often assembled in the mangrove forests called Rừng Sác (Li 2016).

marginalized and often socially denigrated populations living there are displaced; at the same time, the water in the wetlands is also displaced.

When city residents connect peri-urban population displacement to infilling on the margins and then to flooding in central parts of Saigon, they do so based on commonsense understandings about the way social, environmental, and political conditions are entangled in the city's riverine ecology. In 2010, during the same trip in which I started to notice residents speaking more and more about flooding, I too noticed that streets that never flooded in the past were inundated several times during that rainy season. In some of the city's notorious wet-spots, flooding was more intense and lasted longer than it had in previous years, all despite the fact that the city had been furiously working to improve its network of sewers and drainage pipes.

Newspapers intensified the sense of crisis by printing photo spreads of the flooded city. For the previous several years, newspapers had been making dire warnings that the city might someday become "an island" if sea levels rose any higher (e.g., Kiên Cường 2009). The warnings themselves have become something of a seasonal phenomenon in their own right. A torrent of articles about flooding appears in conjunction with every rainy season, and a rising tide of articles increases with cycles of the moon, which regulate the levels of the Saigon River. Whenever heavy rains and high tides coincide, one can be more or less certain that specific spots of the city will flood. Scientists and geographers have connected this to urban growth by using remote sensing data to study changes in surface topography due to urbanization and the rapid increase in concrete-covered "impermeable surfaces" in the city (Tran Thi Van, Bui Thi Thy Y, and Ha Duong Xuan Bao 2015; Tran Thi Van and Ha Duong Xuan Bao 2010). Every day, residents have been making similar connections by simply looking at the city and putting two and two together.

It is an obvious connection to make. In 2010, during my extended research visit to the city, at the very same time that everyone seemed to be talking more and more about the problem of urban flooding, lines of barges were lining up along the Saigon River, pumping sand into an area known for its "low lying land." Similarly, across town, in the actually existing New Urban Zone of Phú Mỹ Hưng, which was also built on "low land," pavement rollers were busy laying another foot of gravel and tarmac over neighborhood streets, which were already sinking after only a few years of use. Even the land developers are open about this—one project developer in Phú Mỹ Hưng told me that regular repaving was part of the original mitigation plan for expected subsidence, and the company's own internal history is even called "Phú Mỹ Hưng—Rising from the Swamps" (Phú Mỹ Hưng 2005). The developers well knew that the New Urban Zone was built on soft, inundated soil and their solution amounted to little more than continuously filling in the swamp. Meanwhile, as developers kept filling in the sinking streets of Phú Mỹ Hưng and the wetlands of Thủ Thiêm, and as the swamps in both places became less swampy, the rest of the city kept flooding.

In the face of these observations, two profoundly different forms of environmental language and practice have emerged among the former residents of Thủ Thiêm and those charged with turning it into a New Urban Zone. The next sections of this chapter compare the ways that the people evicted from Thủ Thiêm lived within and spoke about the environment in Thủ Thiêm with the ways urban designers talk about "ecology." What becomes clear is that the people displaced from Thủ Thiêm to make way for the New Urban Zone project rarely spoke of "the environment" but also consciously organized significant aspects of their lives in relationship to the diurnal tidal landscape. By contrast, the planners of the New Urban Zone regularly evoke the idea that the area has a unique riverine ecology, all while designing a project destined to radically transform that very ecology.

The Eco-logic of the Thủ Thiêm New Urban Zone

Thủ Thiêm is located immediately across the Saigon River from the city's historic core and contemporary central business district. Thủ Thiêm was never empty, but life there was not defined by the terrestrial bias of modern city building. Instead, it followed a model of habitation common in this part of Southeast Asia, where houses were (and still are) built along the course of waterways and on top of filaments or pads of packed earth, and in some cases on stilts right over the water (Harms 2016a). Meanwhile, the softer soils, crisscrossed by rivulets and canals, were used for agriculture, fishing, and waterborne movement. Until the construction of the Thủ Thiêm New Urban Zone began to transform the landscape in 2002, housing largely followed this pattern, hugging the bend of the river, and grouping along canals and two roads built on raised earth. The low-lying land located behind the densely packed neighborhoods was sparsely occupied and only expanded to accommodate homesteads when population pressure demanded it. Land infilling was always incremental and largely done by hand.

Thủ Thiêm residents, who lived just across the river from downtown, were just as urban as anyone else in the city. When I was conducting research there, however, I was often struck by what might be called their unspoken ecological consciousness. By unspoken, I mean that none of the residents I met during my fieldwork ever used terms like "ecology" or "the environment" to describe their lifestyles; nevertheless, by ecological consciousness, I mean that they actually lived their lives in ways that evinced a distinct practical awareness of and attention to environmental factors. For example, people in Thủ Thiêm never used air conditioning. They preferred to maximize breezes and regulate the shade and sun with screens and shading devices. They also regulated their activities and movements into and out of the city according to time of day. Unless otherwise forced by circumstances beyond their control, they tended to wake up early to enjoy the cool mornings, avoided travel during the hottest time of day, rested when the sun was at its peak, and stayed up late into the cool evening. Most houses were shaded by trees that the residents planted themselves. Many houses, if they had enough land, were surrounded by gardens. Those without gardens were surrounded by potted plants. Access to homes was not via oversized roads but by thin, self-built cement pathways shored up by elephant grass and coconut trees. As a result, few people owned personal automobiles, and nearly everyone commuted by motorbike or bicycle. If one took all of these urban practices and typed them up in a bullet-point list, one could write a manifesto of green urbanism that would look something like this:

- Avoid using air conditioning. Keep cool by controlling the shade and capturing the breeze. Don't run errands in the hottest part of the day unless necessary.
- Minimize automobile access. Construct narrow pathways to homes. Confine automobile access to major roadways, and limit traffic.
- Surround yourself with plants. Be sure to combine a mixture of ornamental plants and edible plants—including fruit trees, herbs, water-spinach, and medicinals. Minimize the clearance of natural vegetation, especially along waterways.
- Support mixed-use zoning. Weave shops and housing together into the urban fabric. Encourage the integration of small-scale shop life into residential homes. Minimize commuting by providing daily necessities within walking distance.
- Practice self-contained waste management. All waste produced should be collected and processed by the waste producer.
- Encourage multigenerational co-habitation. As families grow, the house is to be considered a "project," which can be built up, expanded, or subdivided as needed in order to accommodate multiple generations.
- Integrate spiritual life into the community. All neighborhoods should include access to churches, temples, pagodas, shrines, sacred spaces, and other features of spiritual life.
- Promote a vibrant "café society." Mixed-use zoning promotes a proliferation of open-air cafes and sidewalk eateries within walking distance of any household.
- Integrate urban agriculture into everyday urban life. Promote the dissolution of the distinction between country and city.
- Promote "slow food." Encourage long lunches, extended weekend meals with friends in outdoor settings, and other forms of shared alimentary commensality. Enjoy meals cooked at home using fresh, seasonal, local ingredients.

This "manifesto," of course, is a post-facto list constructed out of observations of everyday behavior—an etic perspective compiled by the anthropologist over a period of participant observation. What it shows, however, is that while Thủ Thiêm residents rarely ever spoke explicitly of "the environment," they were very aware of, and sought to maximize the affordances provided by, their ecological setting.

A sense of ecological awareness was even evident during everyday leisure activity in Thủ Thiêm. On many occasions, for example, I would find myself sitting behind a friend's house with groups of men who enjoyed gathering together in a bamboo gazebo the friend had built on stilts above a natural waterway. The gazebo was shaded by a roof made from water coconut palm fronds (*Dừa nước, Nypa fruticans*; also known as Attap, Nipa, or Nipah palm), which were gathered from the area behind his house. Occasionally, while lounging in this setting, some of the men in this group would all of a sudden leap to their feet and then wade out into the ponds that formed amongst the roots of water coconut trees. As they waded through the ponds, they would reach through the reeds and water-spinach, feel around for a moment, and then pull out delicacies like frogs and water snakes. Then they'd climb back up to the gazebo and continue doing whatever they had been doing up until that point, which could be anything—reading the paper, smoking, resting on a hammock, tending to a fighting cock, or making a deal on their cell phone.

One of the best hunters in one such group of friends was locally famous as the chain-smoking, hard-drinking money collector for a smalltime local mafia group. He had multiple "wives," wore gold chains, and rode the fanciest brand of motorbike. However, whenever he came by the gazebo in Thủ Thiêm, he would often head straight to the water's edge, crouching low, stalking like a hunter through the thickets and overgrowth, always knowing where to find another product of Thủ Thiêm's ecological bounty. Most of the other people living in Thủ Thiêm also knew the waterways like the back of their hand. They knew which eddies under which bridges, or behind which clump of water-coconut root, tended to host the greatest number of catfish, and which ponds were the best place to catch ornamental fish. They knew where to find the strongest palm fronds for thatching a roof, where to harvest the plump seeds of water-coconut, when the tides would rise and fall, and so on.

Despite all this, my friends in Thủ Thiêm never seemed to talk about "the environment." Despite their extensive local knowledge of the ecological setting in which they lived, they would casually throw spent cigarette butts and empty cigarette boxes, as well as a wide range of plastic wrappers and baggies, into the bushes and the ponds. Food and plastic scraps were scattered everywhere. The close communion humans in Thủ Thiêm shared with the nonhuman was not a utopian communion with green plants and pristine, untouched nature. Humans also lived in communion with the anthropogenic flotsam and jetsam of their own activity, like plastic snack bags, discarded rubber flip-flops, and, as with so many places in Vietnam today, piles upon piles of plastic baggies, tangled up in the roots of all the fecund undergrowth. On the one hand, the amount of garbage one saw in the area could be read as evidence for how little Thủ Thiêm residents thought or cared about the environment. On the other hand, it also reflected how these areas lacked regular garbage collection. The close entanglement of plastic waste with human life and nature was disconcerting, but it also revealed a way of seeing anthropogenic waste as something entangled in a relationship to the space in which they lived. The area formed, in a way, a complex urban ecosystem, into which waste was integrated rather than simply ejected. Rather than shipping their waste off to distant landfills, Thủ Thiêm residents folded it into their own environment.

Concrete Ecologies

While my friends living in Thủ Thiêm rarely spoke about ecology, the official plans for the Thủ Thiêm New Urban Zone, by contrast, mention ecology and the environment at every opportunity. Curiously, the actors who are most explicit about their celebration of the unique ecology of the southern Vietnamese landscape are precisely the ones designing a project that openly endeavors to fully transform it.

Consider the ideas of Dennis Pieprz, one of the principals for Sasaki Urban Design Associates, and one of the main urban planners involved in developing the master plan for Thủ Thiêm. All the evidence shows that Sasaki is dedicated to designing an ecologically progressive model for Thủ Thiêm, and that the firm truly believes in using the project to help guide the city on a pathway of sustainable development. Pieprz even wrote an academic article in a volume on urban sustainability showcasing the way in which the Thủ Thiêm plan was explicitly designed to respond to ecological threats facing the city. "Today," he writes, Ho Chi Minh City's unique "character and the city's sustainability are threatened by over-sized office and commercial development and population growth." He then laments the fact that "rapid urbanization is also placing significant development pressure on the city's historic core" (Pieprz 2011, 121). To solve this problem, he explains, the city has turned to the Thủ Thiêm site: "Its transformation has represented a momentous opportunity for Ho Chi Minh City to become more sustainable and it has been a major national initiative for the economic development of Vietnam" (Pieprz 2011, 121). Pieprz then explains that "*The Detailed Master Plan and Urban Design Guidelines for Thủ Thiêm New Urban Area* prepared by Sasaki established a comprehensive twenty-year vision for the waterfront peninsula as a dynamic and environmentally sensitive mixed-use urban district, reflecting a uniquely Vietnamese way of life" (Pieprz 2011, 122). Waterways are a central feature of the proposal, and Pieprz insists that the project has tried to "address sustainable urban development challenges facing Ho Chi Minh City" and that project developers have been attempting to work in relationship with the "ecology of the natural systems that so significantly affect life in subtropical South East Asia" (Pieprz 2011, 122, 123).

Sasaki's focus on sustainability is even more clear in some of the diagrams the firm produced for the master plan, which clearly deploy a wide range of cutting-edge technologies and concepts to realize this vision of ecological urbanism. One drawing in a section of the master plan devoted to "sustainability" depicts an architectural cross section of proposed building types planned for the peninsula. Surrounding the drawings of the buildings, a series of text boxes and arrows highlight core sustainability features of the development. These text boxes, when listed out, can be read like a bullet-point list equivalent to the manifesto of green urbanism presented in the previous section. Under the heading "passive strategies," the following features are listed:

- North-South building orientation minimizes West-East exposure, reducing heat gain and maximizing wind circulation
- Shading devices and openable windows further reduce heat gain on W-E facades while allowing cross ventilation
- Building heights strategy maximizes daylight and views
- A combination of green and white roofs reflect solar radiation and minimize heat gain
- Atriums facilitate cross ventilation within towers
- Building podiums with open ground level layout allow for well ventilated public realm
- Greenroofs capture rain water and reduce overall demand and stormwater runoff while providing pleasant exterior spaces
- Recycling waste water on site and using it for flushing and irrigation reduces water demand
- Open spaces designed to detain and filter first-flush pollutants from stormwater and slow runoff
- Canals lead into the **Central Lake** and **Southern Delta**, receiving landscapes in periods of high tide, as well as water filtration systems for water quality

In a section on "integrated sustainability strategies," a diagram highlights the movement of the sun across the sky and points to various flows of water and energy, drawing particular attention to the plan's focus on shadows and wind (Sasaki 2012, 10):

- North-South building orientation minimizes West-East exposure, reducing heat gain
- Secondary Winds: For a hot-humid tropical climate it is critical to allow air movement through the site. The plan responds to secondary wind direction to allow for air movement through streets.
- Primary Winds; Oblique building orientations break the prevailing winds and allow light breezes through the streets.

Clearly, the Sasaki plan has an explicit and highly conscious environmental agenda. Nevertheless, while Sasaki explicitly celebrates the local ecological landscape, all of this attention to the local ecology is founded on a set of fundamental contradictions. First, the Sasaki plan presents this southern ecology, with its specific winds, shadows, water flows, canals, and vegetation as if it is some kind of novel scientific discovery when in fact repetitive speech about the "environment" or "sustainability" serves to displace the very ecology about which it speaks. Indeed, if one looks at the landscape as it is being transformed to realize this vision, it is impossible not to see that this southern ecology, so recently discovered by the Sasaki plan, is actually being covered in concrete. Several new concrete provisioning plants have been built specifically for the Thủ Thiêm project, including two plants run by the Le Phan Cement Corporation that together produce 240 cubic meters of cement per hour for one housing development in Thủ Thiêm (Le Phan Concrete 2015). Additionally, the Saigon RDC concrete batching plant, located under the Thủ Thiêm bridge, is one of the newest of that company's seven plants across the city, which together produce 1,000 cubic meters of cement per hour and over a million cubic meters per year (Saigon RDC n.d.). These concrete plants have been built specifically to carry out the plan designed by Sasaki and subsequent designers. Furthermore, the displacement is not just natural but also social. In the process of replacing an actually existing ecology with an abstract discourse of ecology, the project has required evicting 14,600 households from their homes (Harms 2014). In sum, the displacement of actual ecology by a discourse of ecology emerges from the double process of socionatural displacement that ejects both water and people from Thủ Thiêm.

In addition to Sasaki's 1:2000 scale plan for all of Thủ Thiêm, the 1:500 scale design for the central square by DeSo architectes exhibits a similar tendency to promote "the ecology" of Thủ Thiêm at precisely the moment in which plans to pave it in concrete are being outlined. DeSo architectes is a small French architectural design firm with an agenda that claims to be socially and environmentally conscious, and which devotes a great deal of space in its design proposals trying to reconcile environmental concerns with the demands of the clients (the Vietnamese government, the Ho Chi Minh City People's Committee, and various investors). For example, in a recent revision of their detailed plan, DeSo explains its core philosophy:

> The project is not simply a planning drawing on the surface, it is also the attention to the thicknest [sic] of the soil layer, to the accumulation of time into geography. Our desire is to form new architecture, modern, diversified style, but it must be permanently stand [sic] on the land representing identities of Saigon, in order to create an unique public space. The project holds an important definition is sedimentation: sediment from Saigon river that make the soil and tell the city's history (DeSo 2016, 6).

While cynical readers might dismiss such statements as a form of disingenuous "greenwashing," such a reading does not capture the sincerity expressed by companies like DeSo. As with the Sasaki plan, the DeSo plan exudes a sincere commitment to developing sustainable design solutions, and the architects seem committed to teasing out ecological design principles from their work, even as they face specific constraints.

Such sincere expressions of environmental consciousness among designers involved in projects that radically transform the environment are not unique to Vietnam. In her research among "green architects" in Mumbai, Anne Rademacher has described how architects must grapple with the challenges of bringing green design to life: "despite its socially vital life as aspiration, in practice, good design was inevitably dormant. Even standing at-the-ready, its enactments seemed always almost fully dependent on external political economic and ecological activation" (Rademacher 2017, 164). A similar sense pervades the work of architects like Sasaki and DeSo. On the one hand, their drawings depict vast swaths of engineered space covered in concrete; on the other hand, the plans express sincere commitment to preserving local nature and landscapes. For example, DeSo asserts that their designs seek to "respect and regenerate vegetation areas, cultural elements and local specific practices, cherish creative, dynamic personality of the city" (DeSo 2016, 6). The depiction of nature that emerges is a tangle of aspiration, politics, and transformed human-engineered landscapes.

Conclusion: A Dialectic of Covering and Discovering Nature

Urban designers working on the Thủ Thiêm New Urban Zone in Saigon often speak in the pseudo-scientific language of ecological urbanism to justify a project that requires displacing both a riparian ecosystem and thousands of households. Their urban designs commonly claim to have discovered the very thing their own project is covering up. Conversely, the residents who have been displaced from Thủ Thiêm rarely spoke of their lifestyles as ecologically friendly while in fact many aspects of their lifestyles could be understood as a form of green urbanism. Meanwhile, as the people from Thủ Thiêm were being evicted, everyday residents across Saigon were increasingly complaining about urban flooding, and much of the blame was being directed toward the developers of New Urban Zones who are filling wetlands with impermeable surfaces. Furthermore, while the urban designers deploy the language of ecological science to justify their projects, it turns out that most of the actual science being conducted on flooding, urbanization, and the expansion of impervious surfaces supports the everyday observations Saigon residents have long been making about the effects of urban development projects being built in the city's wetlands.

In contemporary Saigon, when looking at and commenting on the spread of concrete that continues to blanket the city with impermeable surfaces, and while making a causal connection between that spread and the rising incidences of urban flooding, everyday residents engage the science of the concrete—a mode of reason that links causality to socially and materially significant effects in the world around them. In doing so, they also link the stories of socionatural displacement with other major organizing concepts commonly used for understanding Vietnamese social life—dispossession, corruption, political incompetence, greed, and graft. The displacement of water caused by urban construction thus takes on multiple meanings, all of which are connected to urban development and political legitimacy. People are not blind, and they increasingly recognize that urban development was not only displacing water but people too. Tracing these kinds of connections is central to any kind of socionatural analysis, and this chapter's attempt to trace some of these interconnected social, natural, material, and political effects is inspired by the ecologies of urbanism approach developed in this book. Such a mode of analysis, furthermore, resonates with the ways many people in Saigon seek to highlight the connections between different forms of displacement, which can be social, natural, and even discursive. Such a mode of analysis reveals an important link between the social dynamics of New Urban Zones and the story of urban ecologies.

The interconnections noted in this chapter, it is worth noting, are not unidirectional but interact dialectically. Social conceptions of the city impact the ways people understand its ecology, and the material forces of nature transform the way people understand their social place within the city. In this case, the act of building New Urban Zones, and the latent social and ecological processes set in motion by doing so, led people to "discover" ecology at the same time that the building of those zones has set in motion processes destined to forever transform that ecology. The planning statements and advertisements describing these zones constantly remind people about the unique ecological context of Ho Chi Minh City. Building these zones, however, has depended, in large part, on covering that ecology in concrete.

Works Cited

Anand, Nikhil. 2017. *Hydraulic City: Water and the Infrastructures of Citizenship in Mumbai*. Durham, NC: Duke University Press.

Björkman, Lisa. 2015. *Pipe Politics, Contested Waters: Embedded Infrastrucures of Millennial Mumbai* Durham, NC: Duke University Press.

Dasgupta, Susmita, Benoit Laplante, Craig Meisner, David Wheeler, and Jianping Yan. 2007. *The Impact of Sea Level Rise on Developing Countries: A Comparative Analysis*. World Bank (Washington, DC). https://openknowledge.worldbank.org/bitstream/handle/10986/7174/wps4136.pdf.

de Grammont, Lucien. 1863. *Onze Mois de sous-préfecture en Basse-Cochinchine*. Paris: J. Sory, Imprimateur-Editeur.

DeSo. 2016. *Project General Description: 1/500 Architecture & Landscape Master Plan, Central Plaza and River Park, Thu Thiem New Urban Area*. May 2016. DeSo Architectes (N/A).

Dinh Ho Tong Minh, Le Van Trung, and Thuy Le Toan. 2015. "Mapping Ground Subsidence Phenomena in Ho Chi Minh City through the Radar Interferometry Technique Using ALOS PALSAR Data." *Remote Sensing* 7: 8543–62. https://doi.org/doi:10.3390/rs70708543.

Duy, Phan N., Lee Chapman, Miles Tight, Phan N. Linh, and Le V. Thuong. 2018. "Increasing Vulnerability to Floods in New Development Areas: Evidence from Ho Chi Minh City." *International Journal of Climate Change Strategies and Management* 10 (1): 197–212. https://doi.org/doi:10.1108/IJCCSM-12-2016-0169.

Garschagen, Matthias. 2015. "Risky Change? Vietnam's Urban Flood Risk Governance between Climate Dynamics and Transformation." *Pacific Affairs* 88 (3): 599–621.

Harms, Erik. 2011. *Saigon's Edge: On the Margins of Ho Chi Minh City*. Minneapolis: University of Minnesota Press.

Harms, Erik. 2014. "Knowing into Oblivion: Clearing Wastelands and Imagining Emptiness in Vietnamese New Urban Zones." *Singapore Journal of Tropical Geography* 35 (2): 312–27.

Harms, Erik. 2016a. *Luxury and Rubble: Civility and Dispossession in the New Saigon*. Berkeley: University of California Press.

Harms, Erik. 2016b. "Urban Space and Exclusion in Asia." *Annual Review of Anthropology* 45 (October): 45–61.

Kiên Cường. 2009. "TP HCM sẽ thành đảo nếu nước biển dâng thêm 0,5 mét" [HCMC will become an "island" if sea levels rise 0.5m]. *vnexpress*, June 25, 2009.

Kiernan, Ben. 2017. *Viet Nam: A History from Earliest Times to the Present*. New York: Oxford University Press.

Le Phan Concrete. 2015. "Khởi công xây dựng Nhà máy Sản xuất Bê tông tươi phục vụ Dự án Đại Quang Minh." [Groundbreaking for fresh concrete plant serving the Đại Quang Minh project]. Le Phan Construction Co., Ltd. Accessed May 16, 2019. http://www.lephan.com.vn/newsDetail.php?id=265&vnslang=en.

Lévi-Strauss, Claude. 1966. *The Savage Mind*. Chicago: University of Chicago Press.

Li, Kevin. 2016. "Partisan to Sovereign: The Making of the Bình Xuyên in Southern Vietnam, 1945–1948." *The Journal of Vietnamese Studies* 11 (3–4): 140–87. https://doi.org/10.1525/jvs.2016.11.3-4.140.

Marks, Danny. 2015. "The Urban Political Ecology of the 2011 Floods in Bangkok: The Creation of Uneven Vulnerabilities." *Pacific Affairs* 88 (3): 623–51. https://doi.org/10.5509/2015883623.

Miller, Michelle Ann, and Mike Douglass. 2015. "Introduction: Governing Flooding in Asia's Urban Transition." *Pacific Affairs* 88 (3): 499–515. https://doi.org/https://doi.org/10.5509/2015883499.

Nicholls, Robert, Susan Hanson, Celine Herweijer, Nicola Patmore, Stéphane Hallegatte, Jan Corfee-Morlot, Jean Château, and Robert Muir-Wood. 2008. "Ranking Port Cities with High Exposure and Vulnerability to Climate Extremes." *OECD Environment Working Papers*, 1. https://doi.org/10.1787/011766488208.

Phú Mỹ Hưng. 2005. *Vươn lên từ đầm lầy* [Rising from the swamps]. Ho Chi Minh City: Phú Mỹ Hưng Corporation.

Pieprz, Dennis. 2011. "A Landscape Framework for Urban Sustainability: Thu Thiem, Ho Chi Minh City." In *The Ecoedge: Urgent Design Challenges in Building Sustainable Cities*, edited by Esther Charlesworth and Rob Adams, 121–32. London: Taylor and Francis.

Rademacher, Anne. 2011. *Reigning the River: Urban Ecologies and Political Transformation in Kathmandu*. Durham, NC: Duke University Press.

Rademacher, Anne. 2015. "Urban Political Ecology." *Annual Review of Anthropology* 44 (1): 137–52. https://doi.org/doi:10.1146/annurev-anthro-102214-014208.

Rademacher, Anne. 2017. *Building Green: Environmental Architects and the Struggle for Sustainability in Mumbai*. Berkeley: University of California Press.

Rademacher, Anne, and K. Sivaramakrishnan. 2013. "Introduction: Ecologies of Urbanism in India." In *Ecologies of Urbanism in India: Metropolitan Civility and Sustainability*, edited by Anne Rademacher and K. Sivaramakrishnan, 1–41. Hong Kong: Hong Kong University Press.

Saigon RDC. n. d. "We've Got You Covered: Our Batching Plants." Saigon RDC Co., Ltd. Accessed May 16, 2019. https://www.saigonrdc.com/capabilities/.

Sasaki. 2012. "Thu Thiem New Urban Area." Sasaki Associates, Inc. Accessed July 21, 2014. http://www.sasaki.com/project/139/thu-thiem-new-urban-area/.

Storch, Harry, and Nigel K. Downes. 2011. "A Scenario-Based Approach to Assess Ho Chi Minh City's Urban Development Strategies against the Impact of Climate Change." *Cities* 28 (6): 517–26. https://doi.org/https://doi.org/10.1016/j.cities.2011.07.002.

Tran Thi Van, Bui Thi Thy Y, and Ha Duong Xuan Bao. 2015. "Evaluating Change of Surface Topography as Urban Development in the South of Ho Chi Minh City Based on Analyzing Remote Sensing Data." *Vietnam Journal of Earth Sciences* 37 (4): 12. https://doi.org/10.15625/0866-7187/37/4/8302.

Tran Thi Van, and Ha Duong Xuan Bao. 2010. "Study of the Impact of Urban Development on Surface Temperature Using Remote Sensing in Ho Chi Minh City, Southern Vietnam." *Geographical Research* 48 (1): 86–96. https://onlinelibrary.wiley.com/doi/abs/10.1111/j.1745-5871.2009.00607.x.

Truong Chinh, and Vo Nguyen Giap. [1937–1938] 1974. *The Peasant Question*. Translated by Christin Pelzer White. Ithaca, NY: Cornell Univeristy Southeast Asia Program.

Whitington, Jerome. 2019. "Bangkok's New City of Walls." *Anthropology News* (November 15): 1–10. https://anthrosource.onlinelibrary.wiley.com/doi/abs/10.1111/AN.1314.

Yarina, Lizzie. 2018. "Your Sea Wall Won't Save You: Negotiating Rhetorics and Imaginaries of Climate Resilience." *Places Journal*. March. https://placesjournal.org/article/your-sea-wall-wont-save-you.

9
Keeping Pace with the Foodshed in Hangzhou

Caroline Merrifield

The small side road, marked with a sign commemorating the Qianlong Emperor's travels, leads to a stone-paved parking lot bracketed by a stone wall, a moon gate in the center. On the other side is a garden with jewel-box pavilions set amidst manicured trees and water features, traced with a winding path. These are the grounds of the Grange, a chic farm-to-table restaurant in the scenic West Lake District of Hangzhou, the capital of Zhejiang Province. Each pavilion room holds a single table; the restaurant hosts a maximum of ten parties at each lunch and dinner service. All the menus are prix fixe, tailored to the availability of particular ingredients on a particular day, to the number of diners in each party, and to the price bracket set by the host of the meal.

The Grange is a small restaurant, reckoning by the number of customers it serves; but each meal is the product of a large and complex backstage operation. The restaurant staff includes five full-time procurement agents who drive between the Grange and Hangzhou's outlying farming areas each day to purchase fresh, seasonal products. Every ingredient used in the restaurant's kitchens is purchased directly from a smallholder farmer or artisan producer—everything from the chickens and greens and peaches to the soy sauce and pickles and vinegar. The restaurant opened to the public in 2004. At the time, questions of food safety and "sustainability" (生态) were not subjects of public debate and concern in the way they are today. The owner of the restaurant, A Dai, says that he started the Grange as a matter of taste. He wanted to serve dishes with flavors he remembered from childhood, from his grandmother's cooking. To accomplish this he found that he needed to source directly from rural producers using "traditional" production methods. At the Grange, "tradition" is shorthand for a range of long-practiced regional techniques: the use of composts instead of synthetic fertilizers; the cultivation of heritage varieties; hand plowing; intercropping and multicropping; foraging; pickling; drying; fermenting.

Over the years, the Grange has developed a sophisticated supply system. A Dai's maternal uncle—called "Uncle" by the entire staff—is the head of procurement. He is also the key initiator and maintainer of relationships with farmer-suppliers in the restaurant's network, work he describes as "building roads and bridges." Uncle estimates that he has been in contact with several thousand suppliers over the years; I estimate that several hundred are currently active in the restaurant's network. Uncle also coordinates the work of the other procurement agents. Brother Ming is primarily responsible for purchasing poultry and eggs at procurement stops across the northern and western parts of Hangzhou's Yuhang District. Baldy, who lives on the eastern side of the city, travels a dense daily circuit in eastern Yuhang, where he buys fresh fish and vegetables. Mister Huang has a few dedicated stops farther out from the city, in different directions, for fish, bamboo shoots, and craft food products like tofu skin. Wei, the youngest, and the most recent hire, covers the gaps between the others' routes. Uncle himself mostly undertakes "specialty" procurement, obtaining the unusual varieties and intensely seasonal ingredients that are purchased only once, or a few times, each year.

During my research at the Grange, I served as Uncle's informal apprentice and a junior member of the procurement team. As I followed these men on their daily routes, crisscrossing the countryside, I learned that the restaurant's supply system is poised on the edge of a coming crisis. As the city expands, agrarian spaces are pushed to a greater distance; and the future of the fresh ingredients the restaurant relies on is increasingly called into question.

Into the Foodshed

Uncle tells the story like this. For instance, we might be driving to a village in Yuhang District, to the north and west of the central city. We leave the restaurant, the green hills ridged in tea fields, and follow the roads out and around the lake, up to the older residential areas near the Zhejiang University campus, where the street sides are packed with small storefronts. Further out, the streets become progressively less dense, and the high-rises are newer and higher. We pass Xixi Wetlands on the left. *There used to be families, whole villages, in the wetlands*, Uncle says. Before it was a national tourist destination, it was one of the areas where he made his earliest procurement trips. The people were all relocated. Right in there he had a fish supplier; over there was a family that supplied him with eggs and vegetables. He lost contact when they moved. *I remember where everything used to be*, he says. *I remember all the changes*. The wetlands are barely twenty minutes away from the restaurant, depending on traffic. These days, the closest regular procurement spots, stops on Brother Ming's and Baldy's daily routes, are at least another twenty minutes out.

Whenever we go on a drive, Uncle gives a running commentary on the places we pass through. He has one mental map of presences: in this village there are crabs to be bought, and in that village a couple raises pigs, and in another there's an orchard with an heirloom variety of honey peach. He narrates a second map of absences. Where I see new property development, wide roads, and standardized city greenification projects, he sees vanished crops, earthen farmhouses, fishponds, and livestock. The city has been moving, first patchily and then more and more densely, into one of China's richest farming regions (cf. An et al. 2018). Homes and villages join Uncle's register of absences; but so do old varieties and breeds, with their particular flavors and quirks, and plots of planted land, with their particular soils and microclimates. "Peasants" are transformed into "non-peasants" as rural residents exchange their rights to village land for urban residency status; and rural land is administratively transformed into city land, clearing the way for ever-intensifying "development" (Qian 2008).

In their daily journeys, the Grange procurement agents trace out the shape of a region that extends north, into Jiangsu Province; east, to Ningbo City; west, to Jiangxi Province; and south, to Fuzhou. Their accustomed routes take them through Hangzhou's outlying districts and counties, and to areas administered by the neighboring cities of Shaoxing, Huzhou, and Jiaxing. The region defined through the procurers' journeys is both cultural and ecological, encompassing traditional locales of production for the seasonal components of traditional Hangzhou cuisine. I use the word "region" in the sense suggested by K. Sivaramakrishnan and Arun Agrawal in their volume *Regional Modernities* (2003). Region "has a spatial connotation," like "local" and "global,"

> but it seeks to map the space between these binary polar extremes, refuses attempts at identifying it with a specific scale or geographic size, and focuses instead on the need to attend to the social networks and flows that give it particular form and content. (2003, 13)

I term the region defined by procurement a foodshed, borrowing a well-established usage in sustainable food systems scholarship and activism (cf. Horst and Gaolach 2014). The word "foodshed," a play on "watershed," was coined by W. P. Hedden in his 1929 book *How Great Cities Are Fed*. In a now-classic article, published nearly seventy years later, authors Kloppenburg, Hendrickson, and Stevenson (1996) argue for the revival of the term:

> How better to grasp the shape and the unity of something as complex as a food system than to graphically imagine the flow of food into a particular place? . . . [T]he replacement of "water" with "food" does something very important: it connects the cultural ("food") to the natural (". . . shed"). The term "foodshed". . . starts from a premise of the unity of place and people, of nature and society. (1996, 34)

In short: the "flow" of foods is shaped by uneven features of ecological and social terrain alike. Although "foodshed" is now often used simply to mean "the geographic area from which a population derives its food supply" (cf. Peters et al. 2008), Kloppenburg, Hendrickson, and Stevenson adopt the term for its normative punch. From their standpoint in Madison, Wisconsin, in the late 1990s, they see a system increasingly dominated by transnational corporations and industrial agriculture. They caution the reader that such a system will "restructure this marvelously diverse world into a homogenous plane free of physical or social obstacles to the free flow of money and agricultural commodities" (Kloppenburg, Hendrickson, and Stevenson 1996, 35).

In defiance of homogenizing trends, foodsheds are:

> self-reliant, locally or regionally based food systems comprised of diversified farms using sustainable practices to supply fresher, more nutritious foodstuffs to small-scale processors and consumers to whom producers are linked by the bonds of community as well as economy. (Kloppenburg, Hendrickson, and Stevenson 1996, 34)

The authors call for "economic exchanges conditioned by such things as pleasure, friendship, aesthetics, affection, loyalty, justice and reciprocity" (1996, 37). They call for commensality—the sociality of eating together—and for seasonal diets tied to local ecologies. They submit that global capitalism is the ultimate cause of problems in the food system; but that, in the everyday meantime of living with/in capitalism, "people working toward foodshed objectives will need to . . . maintain or create alternatives that will eventually bring substantive change" (1996, 37).[1]

The authors' vision coincides with and differs from procurement at the Grange in illuminating ways. The procurement agents drive out and back, every day, to purchase seasonal ingredients from suppliers who employ labor-intensive "traditional" production techniques. Procurement agents and suppliers relate to each other with affection and loyalty; and the Grange pays high rates for high-quality products. In these respects, the foodshed defined through procurement is the sort of ideal foodshed imagined by Kloppenburg et al. However, the authors also assume a baseline situation in which "conventional" agriculture is large-scale, mechanized, input-intensive industrial monoculture, as has firmly been the case in the United States since the end of the Second World War (Adams 2011; Dimitri, Effland, and Conklin 2005; Fitzgerald 2003). Under these circumstances, "foodshed" is a way of framing emergent possibilities for a better future that might flourish amidst "capitalist ruins." Anna Tsing explains in *The Mushroom at the End of the World* (2015):

> The timber has been cut; the oil has run out; the plantation soil no longer supports crops. The search for assets resumes elsewhere. Thus, simplification for alienation produces ruins, spaces of abandonment for asset production. . . . Still, these places can be lively despite announcements of their death; abandoned asset fields sometimes yield new multispecies and multicultural life. In a global state of precarity, we don't have choices other than looking for life in this ruin. (2015, 6)

US industrial farming, with its severe "high modernist" simplifications (Scott 1998), is predestined for failure and abandonment by capital. Foodshed work, then, is proposed as a strategy for cultivating new life to take its place.

---

1. Here, the authors mean something more like Polanyi's "protective counter-moves" ([1944] 2001, 79) than Gibson-Grahams' "diverse economies." See Gibson-Graham (2008).

Both Tsing and the authors of "Coming in to the Foodshed" (Kloppenburg, Hendrickson, and Stevenson 1996) are attuned to the time of capital's aftermath. In *Imperial Debris*, Ann Stoler (2013) argues that "ruin," the noun, is less helpful than the verb: she is concerned with what is continuing to be produced through the (after)lives of "imperial formations." "Ruin" is a way of indexing the present-day, life-and-death stakes of supposedly past/"post-" colonial histories. Ruin is a political pattern; when visited on a landscape, it reflects and embodies larger political structures. Macro-patterns of ruination, in particular, must be understood with reference to developmentalist state projects (2013, 21). While Stoler's attention is on "colonial pasts and imperial presence" (2013, 7), her argument rhymes well with conditions in China, which is not straightforwardly "capitalist," nor "postcolonial," nor "post-socialist." The foundational, unequal relationship of the countryside to the city in China is an artifact of an historical and ongoing state-led politics that has yielded "relations severed between people and people, and between people and things" (2013, 8).[2]

With Stoler's insistence on *process* in mind, I argue that the temporality of foodsheds is more complex than Tsing's or Kloppenburg's accounts might suggest. A foodshed like that of the Grange is continually worked into being around and through large-scale political and economic formations, as well as within the cyclical, seasonal time of agriculture. However, Stoler joins Tsing in primarily thinking of and with "degraded environments" (2013, 7), "corroded" landscapes (2013, 10), "abandoned . . . plantations" (2013, 20), and "zones of uninhabitable space" (2013, 21). By contrast, the Grange procurement system is intended to recuperate or repair vital elements of agrarian "tradition," whatever piecemeal state they might be in. It operates, quite deliberately, in the space and time before the landscape is degraded, corroded, abandoned. In other words: before the decisive round of ruination commences.

Ruination of what, and for whom? From the perspective of the Grange procurement agents, Hangzhou seems to be steadily contaminating, and then swallowing, the grounds of its own sustenance. The destined spaces of the future city are today's productive fields, orchards, fishponds, and bamboo groves. As they drive, speeding out from the close green hills and onto the elevated expressways, above fields and factories and new four-story houses, the procurement agents chase the moving border of the restaurant's foodshed. Common agricultural products are no longer available in close proximity to the city. The relatively remote farming areas that have increasingly become core procurement areas for the restaurant are at once deep in the past—not-yet "developed"—and deep in the future, as they hold out a promise of as-yet-still-clean water and soil. The restaurant owner, A Dai, says that these farming areas maintain their "original ecological state" (原生态)[3] and are

---

2. See Schneider (2017) on the production of meat and waste in rural China.
3. See Luo's (2018) incisive analysis of the concept of "original ecology," or *yuanshengtai*, which is closely connected to narratives of indigeneity in China.

consequently suitable for producing fresh, flavorful ingredients for use in seasonal recipes. In contrast, lands nearer to the city, even those where agricultural production continues in some form or another, are often too polluted to yield ingredients with their "essential qualities" (本味) intact. (In chapter 2 of this volume, Camille Frazier discusses strikingly similar concerns about near-city production among members of Bengaluru's terrace gardening community.)[4]

The "agrarian" and the "natural"—the "originally ecological"—are not separate categories for my interlocutors, and I follow their conceptual lead. Thinking the "natural" as a "wild" rather than a working landscape smuggles in ontological presumptions that do not obtain here.[5] Longstanding Chinese[6] conceptions of the nonhuman world prioritize human living on and with the land: in this way of thinking, what is most "natural" to human beings is precisely the well-tended agrarian landscape. This observation is far from new: it is very, very old. But in the spirit of this volume, it is worth remembering that agrarian environments are "natural," just as human existence is part of "nature" (Agrawal and Sivaramakrishnan 2000).

In Hangzhou, one mode of human living—urbanism—seems to be killing off this "naturally" human mode of agrarian life. The expansion of Hangzhou City is bringing about the ruin of agrarian nature from the perspective of specific human—agrarian—ends. Rich farmland will be "ruined" for agriculture once it is "developed" for commercial housing or infrastructure construction. Villages may be socially "ruined" as residents are forced to relocate, the land of their homes and fields claimed for new purposes (Chuang 2020). The thriving life of the city threatens the agrarian landscapes of its foodshed(s), and thereby the lives of urban residents. The regional concept of foodshed underscores how "city" and "countryside" cannot, nor should not, be analytically disentangled (Domingos, Sobral, and West 2014; Williams 1973). Rather than existing somehow "outside" the city, the spaces of agrarian nature that feed it are intrinsic to its ongoing metabolic processes (Rademacher and Sivaramakrishnan 2017; Solomon 2016).

In the following section, I visit a nearby procurement location in the restaurant foodshed to explore how the growth of the city can be ruinous—deathly—to agrarian nature. In my conversations with restaurant staff and suppliers, I found both an expectation of coming ruin in agrarian spaces and a feeling that such ruin

---

4. See Camille Frazier, "Putting the Garden Back: Cultivating Life through Urban Gardening in India," chapter 2 in this volume.
5. As Mark Elvin writes in his analysis of early medieval period poetry about Hangzhou's landscapes, "The apparent conflict between a mystical delight in nature, a respect for it, and a religious and artistic sensitivity to it, on the one hand, and a process of determined exploitation and development, on the other, is largely an artifact of our modern perspective" (2004, 367). See Miller, Smyer, and Van Der Veer (2014) on rich and multifarious conceptions of nonhuman nature in Chinese philosophy. For instance, notions of "wildness" (野) in Chinese thought point to states of being beyond culture. However, "wildness" sits within a constellation of concepts relating to the nonhuman world that do not operate according to a Western-style nature-culture binary. (See Johnson [2019] for analysis of a parallel notion of wildness, *theuan*, in Thailand.)
6. I use "Chinese" here as shorthand. It would be more accurate to refer to historical conceptions of "nature" among peoples now known as "Han," living within states and other political formations in the region we now call "China."

would unfold along an arbitrary and unknown timeline, determined elsewhere. As I drove through the city with the procurement agents, I also encountered unexpected (after)lives of agrarian nature in core urban territories. In a second section, I focus on two now-urban places that were recently food-producing areas. The stories of these places raise provocative questions about the changing definition of "nature" itself—and about the conditions that make it quick, dead, or otherwise.

Development(s)

Yunquan Village, in Xihu District's Shuangpu Town, is around forty minutes' drive south of the Grange. You pass through Meijia Wu, a neighboring tea village, and alongside the faux fortifications of a Song Dynasty–themed amusement park. You pass waves of new buildings and new commercial complexes, and enter the interstitial scrub of tiny hotels, carwashes, and shops selling toilets and fixtures. You make a right at an Anhui-cuisine diner, and drive through clusters of tiled three-story farmhouses, all crammed together; and then you arrive at the last plot of real West Lake water shield, or *chuncai* (莼菜).

Chuncai, a leafy water plant that is eaten as a vegetable, is iconic of Hangzhou, and of the larger Jiangnan region to which it belongs (Swislocki 2009). *Chuncai* has a short harvest period of about twenty days in the spring—or rather, the Grange only sources it for that long. The plants continue producing new leaves for half a year, but the later crops are sold for much cheaper prices, either for use in other restaurants, or to be jarred or canned. According to A Dai, the owner of the Grange, "A lot of so-called gourmands have never eaten the real thing." (Other outlets present large leaves and stems as *chuncai*, when only the small, fresh leaves should be used.) At the Grange, the tenderest new leaves are served in a delicate soup with pale slivers of perch. *Chuncai* is covered in a clear, protein-heavy, mucilaginous layer, which gives it its surprising texture. It has a clean taste—a green grace note—but it is mostly a slippery feeling on the tongue and in the throat.

I visited Master Ge and his wife, Auntie Mao, the Grange's *chuncai* suppliers, on a few occasions in mid-April and early May 2015. They, and their families, have been growing *chuncai* in the area for a long time. Ge, who is now around sixty, remembers that *chuncai* was a major crop in the village when he was a child. As a child, he knew a man in the village, already in his eighties then, who remembered *chuncai* as a crop from *his* childhood. The land in the couple's village is well-suited to *chuncai* cultivation because the crop requires a supply of clean, cold water, which Ge and Mao obtain from a nearby mountain spring. They grow *chuncai* in a flooded field. Auntie Mao harvests the crop by lying face-down in a small wooden boat, slowly paddling her way around the field with her hands, picking the smallest leaf buds from the surface of the water. Back at home, Ge and Mao pour the leaves out on a clean plastic tarp, sorting out smaller leaves to sell to the Grange, and larger leaves for sale elsewhere.

Master Ge and Auntie Mao have the only remaining area of *chuncai* cultivation in their village, and indeed, in all of West Lake District, as far as they know. Holding onto their land may prove difficult. Preparatory constructions for the new subway line already loom on all sides. If they have to give up the *chuncai* plot, they might try to secure land closer to the mountain, where new "development" (开发) is less imminent; but even if they are able to access water there, the soil will not be as good. During my visits, Ge and Mao were preoccupied with home renovations in preparation for their son's wedding. The couple told me that they were spending about eight *wan* (80,000 RMB, about US$12,800) to refurbish the second floor of their house, despite their strong suspicion that the city government would requisition additional village land over the next few years. They were worried about what might happen to their home; and they worried that they would be poorly compensated for their losses.

A Dai sees *chuncai* as emblematic of far-reaching challenges in the restaurant's foodshed. Master Ge and Auntie Mao may keep producing high-quality *chuncai* for another five years, and the Grange will serve *chuncai* for those five years. After that: a blank. *Who can say?* For A Dai, *chuncai* is properly—culturally, historically—a product of the West Lake area. It is a product of certain soil, certain water, and certain forms of knowledge and skill. Subway construction would take a considerable toll on the biophysical landscape. Scattered studies on changing indicators of environmental quality in the district and the larger region point to the likely consequences of heavy construction and traffic, including increased air pollution and heavy metal deposition in the soil (cf. Bai et al. 2016; Shi et al. 2010; Zhao et al. 2015). If, following expropriation and development, the land can no longer be used to cultivate what it was meant to cultivate, in (agri)cultural terms, then it is "ruined," from a certain human point of view. Significant cultural practices, including culinary practices, may be "ruined" along with it. When the metro takes Master Ge's land, it will take the authentic grounds of production for something that may never grow in the same way, with the same flavor, elsewhere.

Master Ge and Auntie Mao, like many of the farmers in the Grange procurement network, are relatively well-off. They make a good income from the combination of things they grow for sale in different seasons. They like to manage their own time, to "do business" for themselves. Their comfortable livelihood is jeopardized by a multifaceted state program of accumulation. In areas close to the central city, rural land is expropriated, and the city expands—first administratively, and then substantively, as new residential and commercial complexes are constructed on farmland (Ho 2005; Ye 2015). In areas more distant from the central city, the government practices abandonment, withdrawing schools and social services, and coercing families to relocate, always farther down the mountain, closer to the town (cf. Wu 2019).

In *The Mushroom at the End of the World*, Anna Tsing asks us to look for life in the ruins, to find hope in "what manages to live despite capitalism" (2015, viii), and in the "possibilities of coexistence within environmental disturbance" (2015, 4). But facing down the coming of ruin to as-yet-unruined places can be an intensely unsettling prospect for the people who dwell[7] there. The potential "ruin" of the *chuncai* field is not only a matter of ecological disturbance: it will bring an irredeemable loss of terroir; a grave injury to the integrity of regional cuisine; and a breach in (agri) cultural knowledge, if Ge and Mao leave farming, and if no one learns from them how to cultivate *chuncai*.

Nature's After-lives in the City

A 1989 public policy book on the problems of urban development in Chinese cities, edited by Gu Yingchun, a scholar affiliated with the Zhejiang Academy of Social Sciences, includes a section on Hangzhou's West Lake County, the administrative precursor to today's West Lake District. At the time, the city government was already contemplating the West Lake area's future as a scenic tourist district; but the growing prosperity of the area's villages was beginning to pose challenges for this vision. The author of this section bemoans the increasingly bustling and "city-like" character of West Lake area villages:

> Aside from closely crowded farmhouses, there have also appeared not a few buildings for industrial, commercial, hospitality, and service use. . . . With the existing strain on useable land, this has made villages even more dense and disorderly, completely taking leave of the tea-producing villages' original bucolic charm. (Gu 1989, 314)

The author repeatedly points out that the ostensibly rural residents of the scenic areas are mostly eating "commercial grain" from the state-controlled supply, as city residents do, and recommends that the city administer these citified peasants and villages as urban people and places. One village in particular, Sandy Banks Village, is proposed as a test case for this kind of transition: by 1989, a large portion of village land had already been expropriated by the city government. This village was the childhood home of Uncle, the head of procurement at the Grange.

Today, Uncle lives in a neighborhood near the Zhejiang University campus, a short, shady walk from the quiet northern side of West Lake. The neighborhood is tucked outside of Hangzhou's bustling central tourist areas, which cluster along the eastern side of the lake. To the south and west of the area, the map is green with parks, gardens, and temple grounds. To the north, there are residential blocks, new shopping malls, and office buildings. Uncle has told me many stories about his childhood: catching frogs and fish in West Lake, grazing sheep near the water, going out into the village tea fields for "farming class" in elementary school. Nonetheless,

---

7. Throughout, I use "dwell" in Ingold's (2000) sense—not as "mere occupancy" (2000, 150), but as immersive, patterned investments in place-making and person-making, within a particular socio-ecological context.

it took a long while before I really considered Uncle's neighborhood, with its clusters of multistory homes offset from a busy main road, and realized: *this* is Sandy Banks.

In Uncle's memory, the village was removed from the city proper; it was located in "an area where city and countryside join together" (城乡接合部).[8] Sandy Banks has now become an "urban village":[9] land retained under the ownership of a village collective but surrounded by the city.[10] Hangzhou was among the first places in the country to introduce a system of "rural collective retained land," in 1999, and Sandy Banks was the first place in Hangzhou where this system was implemented.[11] In this arrangement, the city government claims a particular area of rural land controlled by a village collective for the "public good." A portion[12] of the expropriated land is set aside by the city for the continued use of the collective, either for "secondary or tertiary" industries, or for housing for displaced residents, who gain urban residency status. In the case of Sandy Banks, the village committee invested in the retained land, building two hotels and a parking lot. These businesses were then "contracted out" to be managed, and the village created a share system to distribute profits from the ventures among village residents.[13]

Sandy Banks remains a source of income for members of its collective and a focal point for ongoing relationships between former village residents and their families. The conditions of the collective land tenure arrangement mean that people who grew up together in the village are still neighbors to this day. In addition to living together, they may also work together. Many of the current senior employees at the Grange are neighbors and friends from A Dai's childhood. The urban policy book from 1989 argues that city practices—city ways of living—are killing West Lake villages *qua* villages. These category-blurring places must be regularized through incorporation in the city. Sandy Banks is an early example of the city's

---

8. See Qian (2014) for a history of Hangzhou's urban development in the decades before Reform. The area around Sandy Banks was not always cultivated in the same way; intensive horticultural production there began during the Cultural Revolution, under an ideological drive to make cities "productive."
9. Siu (2007) renders this term, "城中村," as "urban village enclave." The enclaves she studies in Guangzhou present a striking contrast to this example from Hangzhou. While Sandy Banks is now indistinguishable from the surrounding city, Siu describes a situation in which residents of urban village enclaves must defend their "rurality" to retain control over collective resources. Wang et al. (2014) present the term "scenic area village" (景中村)—or "village amidst the scenery"—to emphasize the somewhat unusual circumstances of Hangzhou's earliest "urban villages."
10. Land not owned by a village collective is owned by the state, although it is controlled by nested administrative sub-units. Land under the administration of Hangzhou City is owned by the state, but planning decisions and land-use regulations affecting this land may be made at various levels, from the city district to the province.
11. The term is "征地留用地制度." See Hangzhou World Cultural Heritage Monitoring and Management Center (2014).
12. Qian (2015) writes that, initially, the city of Hangzhou granted 10 percent of expropriated land back to villagers for purposes of "collective economy." In 2001, the city revised the policy so that rural collectives were granted access to land based on "market values."
13. Uncle told me that he takes in about three *wan* (30,000 RMB) per year from the share system. The written sources I consulted show that most shareholders make between 15,000 and 20,000 RMB per year. See also Hangzhou World Cultural Heritage Monitoring and Management Center (2014, 78). This text praises Hangzhou's retained land policy for creating a stable, long-term income source for rural residents, in contrast to land expropriation arrangements in which rural residents are compensated with a lump-sum payment.

post-reform expansion into near-city agrarian environments; it stands now as a case of agrarian after-life, or ongoing liveliness. When "comprehended as something processual" (Agrawal and Sivaramakrishnan 2000, 6), Sandy Banks appears as a sustained agrarian presence, despite its conversion into a piece of central Hangzhou with hotels and restaurants and bus stops and shady sidewalks. The live social and economic bonds conditioned by this continuing existence keep it lively, in turn.

The history of Sandy Banks shows that spaces of agrarian nature have social lives, as much as ecological ones. What keeps Sandy Banks alive, if no longer under cultivation, is a specific land tenure arrangement in which local residents have retained some control over the development trajectory of their village. But Sandy Banks also stands as an exception to now-dominant modes of state-led urban development and land acquisition in China (Hsing 2010; Ren 2020). The role of the state matters tremendously; and, as Master Ge and Auntie Mao, the *chuncai* farmers, suggested to me, the state's plans are made far out of sight of ordinary rural residents.

If Sandy Banks shows that urban expansion might not "kill" the social lives of particular agrarian environments, the case of Hangzhou's Xixi Wetland provides an instructive contrast. The wetland, which once housed several farming villages, likewise transformed as Hangzhou expanded. Yet rather than turning into streets and shops and residences, it was repurposed as a locus of "pure nature" within the city, upending the lives of rural residents with little control over their own land.

Xixi Wetland was established as a national park by the National Forestry Administration on February 2, 2005, World Wetlands Day. It opened to the public in May of that year.[14] The wetland is now marketed as a refuge for "ecology" within the city. From the park website:

> Xixi's core emphasis is on ecology. To strengthen ecological protection . . . five large ecological protection and restoration areas have been established within the park . . . Xixi is also "heaven for birds," and there are many places to watch birds within the park, allowing visitors to experience the majestic vista of flocks of birds in joyful flight. (Xixi National Wetland Park 2018, my translation)

In contemporary descriptions, Xixi Wetland, which is situated partly within West Lake District and partly in the neighboring Yuhang District, is held in contrast with West Lake itself.[15] While the lake is presented as "cultural"—man-made,[16] poetic,

---

14. Official plans for the park, which was initially designated a "cultural ecological tourism area" (文化生态旅游区), were published in November 2001 as part of a comprehensive development strategy for western Hangzhou.
15. The "cultural landscape" of West Lake was inscribed as a UNESCO World Heritage site in 2011. This short description from the UNESCO website is typical of contemporary tourism materials about the lake: "Its beauty has been celebrated by writers and artists since the Tang Dynasty (AD 618–907). In order to make it more beautiful, its islands, causeways and the lower slopes of its hills have been "improved" by the addition of numerous temples, pagodas, pavilions, gardens and ornamental trees which merge with farmed landscape. . . . Since the Southern Song Dynasty (thirteenth century) ten poetically named scenic places have been identified as embodying idealised, classic landscapes—that manifest the perfect fusion between man and nature." See UNESCO (2018).
16. West Lake was originally part of the Qiantang River. Over many generations of sedimentation and large-scale

painterly, historic—the wetland is positioned as a reservoir of "wild" nature and "green" habitat for nonhuman species, and as the "lungs" or "kidneys" of the city.

A 2006 paper by Chen Jiuhe, a land-use planning scholar at Hangzhou Normal University, exemplifies this way of thinking about Xixi Wetland. Throughout, Chen argues that "green landscape design" is integral to the park's capacity to "protect ecology." By carefully choreographing tourists' experiences of the wetland through constructions like paths, signs, and observation areas, the park's planners educate human visitors about the wetland's "nature" even as they shield nonhuman nature from overexposure to humans. While the park complex includes hotels, shops, and restaurants, these are presented as means to channel the flow of visitors through the park, simultaneously entertaining and "strengthening people's awareness of environmental protection" (2006, 75). The park is a "biodiverse ecological system" to be stewarded, even as tourists visit for purposes of leisure and sightseeing. Chen writes: "[H]ere, waterways interlace; there are streams and islets, one upon the other; the reeds grow lush and thick; it is a place for water birds to roost" (2006, 72). Nature, he concludes, is humanity's greatest "source of wealth" (2006, 75); yet the "moral ecology" (Rademacher 2018) exemplified in his account would exclude humans from a truly "restored" wetland.

In a similar vein, Wang Hongxin, a professor of public administration at Beijing Normal University, uses Xixi Wetland as a case study in a 2009 book on Chinese land policy. He opens with an image of a tranquil waterway. Plants and animals are the masters of this domain; but this was not always so. He writes:

> Can you imagine it? A few years ago, humans were the "rulers" here; crowded villages and a dense population were placing excessive demands on the natural environment. In these villages, pig husbandry made up nearly 70% of the total output value of agriculture; at its peak, [the industry] had 20,000 pigs in enclosures, and the resulting pollution vastly exceeded the bearing capacity of the wetland. Fish were being farmed too densely, causing serious eutrophication; there were chaotic houses and dwellings, piles of garbage everywhere. . . . A layer of grime had settled on Xixi, with its more than eighteen hundred years of history. (Wang 2009, 231, my translation)

In 2004, when work began on the program to restore Xixi to its "original ecological state," over a thousand people had to be displaced from their land (see also Hu and Gao 2006). Before Chinese New Year, 2005, many families from the wetland villages were resettled in newly constructed apartment blocks to the north of the park. According to Wang's case study, the city, district, and county-level authorities jointly authorized additional compensation "beyond the usual standard" to hasten the resettlement process. Displaced farmers were also included in the municipal workers' retirement insurance system. Wang assures his readers that people who

---

dredging, it became a freshwater "lake" just outside Hangzhou's old city walls. The lake was incorporated into Hangzhou as a central feature of the city quite recently, during the Republican Period. See Wang Liping (1999).

had to leave the wetland "truly were able to live content and harmonious 'urban lives'" (2009, 232). (As Kasia Paprocki[17] writes in chapter 3, it is worth asking how and when visions of well-functioning "nature" "require the active destruction of rural futures in order to forge new, resilient, and prosperous urban ones.")

My introduction to Xixi was as a tourist. When I first visited the Grange in 2011, A Dai took me out for a day of sightseeing. We rode a boat amongst the reeds, climbed a two-story pavilion to look out across the trees and waterways, and snapped photos in the gardens. The following summer, when I began riding along with Uncle on procurement trips, I learned that the wetland had been the chief procurement area for the Grange in the restaurant's early years. In Uncle's telling, he began sourcing ingredients for the restaurant from the places he most associated with those ingredients—places that had been "known for" certain products over time. The villages in the wetland had been a "known" near-city production ground for poultry, fish, vegetables, and many other essential items.

In a 2015 paper on Xixi, Ma Zhihui, a historical urban geographer at Hangzhou Normal University, emphasizes that the wetland had been the site of farming activity as early as the Neolithic Liangzhu culture. For Ma, the story of the Xixi area is one of humans continually remaking a tricky, flood-prone landscape to suit their (agri)cultural purposes. By the time of the Southern Song, when Hangzhou—then called Lin'an—became the imperial capital, Xixi Wetland was already well-integrated with the city as an outlying commercial district. In addition to intensive farming systems incorporating fish, mulberry, and sericulture, farmers in the area grew tea, bamboo, persimmons, water caltrop, and lotus root. The area was famous for its local variety of Chinese plum, much-coveted for use in bonsai. A network of bustling rural markets across the wetland offered local specialty products like fermented sauces and liquor.

While other authors describe Xixi as a "roosting place" for birds (e.g., Wang 2009, 232), Ma deliberately describes it as a roost for people. He writes that Xixi has "rich biological resources that provided favorable conditions for humankind to roost, and for agricultural development" (Ma 2015, 87, my translation). The depopulation of the wetland in 2004–2005 was framed as a necessary condition for its "recovery"; but as Ma shows, Xixi had been a populated agrarian area since the beginning of recorded history in the region. The wetland's "recovery" also marked the end of food production within close reach of central Hangzhou.

Crucially, the "nature" of Xixi is distinct from the "nature" of ordinary greenification practices in city parks, or along municipal sidewalks and roads. Such plantings are additive and decorative, "beautifying" the city. The wetland park is presented as a reinstatement of wildness in something that was once wild, thereby re-enlivening it. Human dwelling and farming practices in the wetland were cited as evidence of the wetland's coming demise, unless radical steps were taken—much as

---

17. See Kasia Paprocki, "The Village at the End of the World: Ecologies of Urbanism in Climate Crisis Imaginaries," chapter 3 in this volume.

the "urban" practices of Sandy Banks' residents were cited as evidence of its properly nonrural status. But in making "wildness" live in the wetland, the planners of the park quelled a rich social complex. There is a strange deathliness to this form of nature, though it is supposed to be revivified and revivifying. This is not to diminish the effect that twenty thousand hogs must have been having on the wetland ecosystem: human projects can be a brutal burden to the lands in which we live. But the history of Xixi Wetland demonstrates how rigid definitions of "nature"—as a roost for birds, but never for humans—can stymie livable, dwellable accommodations between people and the cultivated landscapes that sustain them.

Conclusion: Waste/Fallow

In these three places—the *chuncai* field, the urban village, and the wetland—I find complex relationships between the life of the city and the lives of food-producing environments. The notion of "ruin(ation)" is ultimately not capacious enough for the intersecting processes I see in Grange foodshed: the deaths—or coming deaths—of spaces of agrarian nature, the strange deathliness of "pure" nature, and the persistent liveliness of long-ago-urbanized agrarian environments. In these cases, life and death are woven together. They are matters of perspective, unstable categories of being.

As this volume emphasizes, "nature" and "the urban" are likewise unstable categories, impossible to separate with clean analytical distinctions. The Hangzhou foodshed I consider here is constituted through social and ecological linkages between spaces of agrarian nature and the urban core. The city—understood as something processual, understood in regional terms—encompasses its ostensibly "rural" hinterlands. The remote farming villages supplying eggs and sweet potatoes to the Grange are essential to the "life" of the city. At the same time, the city's "life," in processes of land appropriation and "development," may be planting the seeds of its eventual "death." The continued liveliness of agrarian nature, somewhere, remains a necessary condition of life for city residents.

As the Grange procurement agents envision the future of the foodshed, they anticipate a few more good years of sourcing products like *chuncai*. But then, that blank: a profound uncertainty about how fresh, seasonal food will continue to flow into the restaurant. The restaurant's procurement agents may not be able to drive far or fast enough, at some point in the future, to bring back food at all; but even now, the various sources of food, their qualities and identities and essences, are changing irrevocably.

Not by way of an answer, but in the spirit of a question, I introduce a Chinese term, something both capacious and shifting. *Huang* (荒) is used in reference to land that has never been cultivated, or arable land that has fallen into disuse (荒地). Land abandoned by its inhabitants, or left fallow too long, can degenerate into the condition of *huang* (荒掉, *huangdiao*). Marginal lands, borderlands, and frontiers are often described as *huang*. The category of *huang* is defined with a thoroughly human-centric (and specifically Han-centric) sensibility: it contains the idea that humans are rightly enmeshed in the nonhuman world through agrarian lifeways.[18]

In English, the concept of "wasteland" denotes a landscape that has yet to be "brought into value" (Cronon 1996; Irvine 2017). In this sense, *huang* overlaps with "wasteland" insofar as its orienting value(s) are agricultural. However, in contemporary Anglophone scholarship, "wasteland" is often used in a sense similar to Tsing's "ruins," or Stoler's, to name scenes of nonhuman nature that appear as other than "wild" or "pristine": the wrecked and weedy construction site (Schwenkel 2013); the drainage ditch (Irvine 2017); the landfill (Hoag, Bertoni, and Bubandt 2018). For geographer Matthew Gandy (2013), for instance, such "urban wastelands," spaces of "spontaneous nature" within the city, betray the incompleteness of urban plans and belie essentialization of "the urban." In them,

> the sense of nature as active, dynamic, and constitutive of the cultural and material characteristics of urban space reveals the metropolis to be both unfixable and to a significant degree unknowable. (Gandy 2013, 1302)

These insights are well-taken. Yet, unlike "wasteland," *huang* operates in a cultural context that does not share the same configuration of assumptions around the concept of "nature" as variously wild/pure or used/fallen (Cronon 1996). *Huang* is a status that landscapes can move into and out of. Humans remake *huang* lands for human purposes; they cut down trees, cut terraced fields into hillsides, plant crops, build houses, and chase the wild elephants far away (Elvin 2004). But fields and terraces and farmhouses can return to *huang* once again, if humans stop pursuing agrarian activities there. *Huang* is a terrain of agrarian potential.

Huang, a designation that figures prominently in long Chinese histories of state land reclamation, military campaigns, and colonization, is not innocent of the wreckage that human habitation and resource exploitation can bring to particular landscapes (cf. Purdue 1987; Marks 2017; Muscolino 2010; Yeh 2009). Indeed, *huang*'s genealogy highlights the dangers posed by human projects, especially state-centric and colonial projects, to the ongoing liveliness of "nature," in both social and biophysical terms. Yet *huang* also, equally, signals the desirability of patterns of human dwelling that sustain collectives of human and nonhuman life—agrarian ecologies—in the long run.

I borrow the idea of *huang* to focus attention on key social dimensions of "natural" liveliness. In Hangzhou, the local government couches rural land acquisition and "development" in very similar terms to preparing uncultivated soil for planting: the land will be "opened up" (开发), remade. Yet most "developed" urban land can never revert to *huang*, except on a beyond-human timescale. In this light, I read that blank I encountered among the Grange procurement agents—*in five*

---

18. Harms discusses the analogous concept in Vietnam, and its attendant "narrative structure of civilizational advancement" (2014, 312).

years, who can say—as an analytical clue. The key question for the life or death of Hangzhou's foodshed(s) is about the extent to which urban "development" precludes a shift to other forms of human habitation. When and where do landscapes reach the point of no return *to cultivation*? The concept of *huang* marks out a social and ecological threshold in the life of the city—and a culinary one, as well.

In Sandy Banks, the social complex of the village lives on; but no rice is grown in the area anymore. None of the orchards remain. No one I know eats fish from the lake. Yet Uncle still tends a small backyard vegetable garden in a space carved out of the scrubby hill behind his house, a "[gap] of the city" (Sugimoto, this volume).[19] The garden is not exactly legal: only the land his house stands on is truly his. The hill is *huang* that has been reclaimed for cultivation, for good tomatoes and peppers and greens.

Works Cited

Adams, Jane, ed. 2011. *Fighting for the Farm: Rural America Transformed*. Philadelphia: University of Pennsylvania Press.

Agrawal, Arun, and K. Sivaramakrishnan. 2000. *Agrarian Environments: Resources, Representations, and Rule in India*. Durham, NC: Duke University Press.

An, Yuqing Jin Yeu Tsou, Kapo Wong, Yuanzhi Zhang, Dawei Liu, and Yu Li. 2018. "Detecting Land Use Changes in a Rapidly Developing City during 1990–2017 Using Satellite Imagery: A Case Study in Hangzhou Urban Area, China." *Sustainability* 10 (9): 3303.

Bai, Yang, Min Wang, Chi Peng, and Juha M. Alatalo. 2016. "Impacts of Urbanization on the Distribution of Heavy Metals in Soils along the Huangpu River, the Drinking Water Source for Shanghai." *Environmental Science and Pollution Research International* 23 (6): 5222–31.

Cronon, William. 1996. "The Trouble with Wilderness: Or, Getting Back to the Wrong Nature." *Environmental History* 1 (1): 7–28.

Chen, Jiuhe. 2006. "*Hangzhou Xixi Guojia Shidi Gongyuan Shengtai Lvyou Jingguan Lvse Sheji*" [Green design for Hangzhou Xixi National Wetland Park ecotourism scenery]. *Arreal Research and Development* 25 (5): 72–75.

Chuang, Julia. 2020. *Beneath the China Boom: Labor, Citizenship, and the Making of a Rural Land Market*. Oakland: University of California Press.

Dimitri, Carolyn, Anne Effland, and Neilson Conklin. 2005. "The 20th Century Transformation of U.S. Agriculture and Farm Policy." *USDA Economic Research Service Economic Information Bulletin* 3, 14 pages.

Domingos, Nuno, Jose Manuel Sobral, and Harry G. West, eds. 2014. *Food Between the Country and the City: Ethnographies of a Changing Global Foodscape*. London: Bloomsbury Academic.

Elvin, Mark. 2004. *The Retreat of the Elephants: An Environmental History of China*. New Haven, CT: Yale University Press.

---

19. See Tomonori Sugimoto, "The Death and Life of Urban Ecological Commons in Taipei," chapter 6 in this volume.

Fitzgerald, Deborah. 2003. *Every Farm a Factory: The Industrial Ideal in American Agriculture.* New Haven, CT: Yale University Press.
Gandy, Matthew. 2013. "Marginalia: Aesthetics, Ecology, and Urban Wastelands." *Annals of the Association of American Geographers* 103 (6): 1301–16.
Gibson-Graham, J. K. 2008. "Diverse Economies: Performative Practices for 'Other Worlds.'" *Progress in Human Geography* 32 (5): 613–63.
Gu, Yingchun, ed. 1989. *Zhongguo de Chengshi 'Bing'* [China's city "disease"]. Beijing: China International Radio Press.
Hangzhou World Cultural Heritage Monitoring and Management Center, Hangzhou City Planning Design Research Institute. 2014. *Chuancheng yu Gongsheng: Zhongguo Shijie Wenhua Yichan yu Shequ Fazhan Yanjiu* [Tradition and symbiosis: Research on China's world cultural heritage and community development]. Beijing: Cultural Relics Press.
Harms, Erik. 2014. "Knowing into Oblivion: Clearing Wastelands and Imagining Emptiness in Vietnamese New Urban Zones." *Singapore Journal of Tropical Geography* 35: 312–27.
Hedden, W. P. 1929. *How Great Cities Are Fed.* Boston: D. C. Heath and Company.
Ho, Peter, 2005. *Institutions in Transition: Land Ownership, Property Rights, and Social Conflict in China.* Oxford: Oxford University Press.
Hoag, Colin, Filippo Bertoni, and Nils Bubandt. 2018. "Wasteland Ecologies: Undomestication and Multispecies Gains on an Anthropocene Dumping Ground." *Journal of Ethnobiology* 38 (1): 88–104.
Horst, Megan, and Brad Gaolach. 2014. "The Potential of Local Food Systems in North America: A Review of Foodshed Analyses." *Renewable Agriculture and Food Systems* 30 (5): 399–407.
Hsing, You-tien. 2010. *The Great Urban Transformation: Politics of Land and Property in China.* New York: Oxford University Press.
Hu, Guojun, and Gao Aiqun. 2006. "Changsanjiao Diqu Chengshi Shidi Baohu yu Keichixu Fazhan—Yi Hangzhou Xixi Shidi Baohu wei Li" [Yangtze River Delta Region urban wetland preservation and sustainable development: The case of Hangzhou's Xixi Wetland]. In *Xunhuan Jingji Lilun yu Shijian: Changsanjiao Xunhuan Jingji Luntan ji 2006 Nian Anhui Boshi Keji Luntan Lunwen Ji* [Cyclical economy theory and practice: Collected papers from the Yangtze River Delta Cyclical Economy Roundtable and the 2006 Anhui Doctoral Science and Technology Roundtable], edited by Li Kunsen and Wei Chong, 152–55. Hefei: Anhui University Press.
Ingold, Tim. 2000. *The Perception of the Environment: Essays on Livelihood, Dwelling, and Skill.* New York: Routledge.
Irvine, Richard D. G. 2017. "Anthropocene East Anglia." *The Sociological Review Monographs* 65 (1): 154–70.
Johnson, Andrew Alan. 2019. "Pa Theuan." In *An Ecotopian Lexicon*, edited by Matthew Schneider-Mayerson and Brent Ryan Bellamy, 187–93. Minneapolis: University of Minnesota Press.
Kloppenburg, Jack, John Hendrickson, and G. W. Stevenson. 1996. "Coming in to the Foodshed." *Agriculture and Human Values* 13 (3): 33–42.
Luo, Yu. 2018. "An Alternative to the 'Indigenous' in Early Twenty-First-Century China: Guizhou's Branding of *Yuanshengtai.*" *Modern China* 44 (1): 68–102.
Ma, Zhihui. 2015. "Gudai Hangzhou Xixi Shidi de Nongye Dili Yanjiu" [Agricultural geography research on ancient Hangzhou's Xixi Wetland]. *Historical Geography* 5: 87–91.

Marks, Robert B. 2017 *China: An Environmental History, Second Edition*. New York: Rowman & Littlefield Publishers.

Miller, James, Dan Smyer Yu, and Peter Van Der Veer, eds. 2014. *Religion and Ecological Sustainability in China*. New York: Routledge.

Muscolino, Micah S. 2010. "Refugees, Land Reclamation, and Militarized Landscapes in Wartime China: Huanglongshan, Shaanxi, 1937–1945." *The Journal of Asian Studies* 69 (2): 453–78.

Peters, Christian J., Nelson L. Bills, Jennifer L. Wilkins, and Gary W. Fick. 2008. "Foodshed Analysis and Its Relevance to Sustainability." *Renewable Agriculture and Food Systems* 24 (1): 1–7.

Polanyi, Karl. (1944) 2001. *The Great Transformation: The Political and Economic Origins of Our Time*. Boston: Beacon Press.

Purdue, Peter. 1987. *Exhausting the Earth: State and Peasant in Hunan, 1500–1850*. Cambridge, MA: Harvard University Press.

Qian, Zhu. 2008. "Evidence from Hangzhou's Urban Land Reform: Evolution, Structure, Constraints and Prospects." *Habitat International* 32: 494–511.

Qian, Zhu. 2014. "China's Pre-reform Urban Transformation: The Case of Hangzhou during the Cultural Revolution (1966–1976)." *International Development Planning Review* 36 (2): 181.

Qian, Zhu. 2015. "Land Acquisition Compensation in Post-Reform China: Evolution, Structure and Challenges in Hangzhou." *Land Use Policy* 46: 250–57.

Rademacher, Anne. 2018. *Building Green: Environmental Architects and the Struggle for Sustainability in Mumbai*. Oakland: University of California Press.

Rademacher, Anne, and K. Sivaramakrishnan. 2017. "Introduction: Places of Nature in Asian Cities and Towns." In *Places of Nature in Ecologies of Urbanism*, edited by Anne Rademacher and K. Sivaramakrishnan, 1–26. Hong Kong: Hong Kong University Press.

Ren, Xuefei. 2020. *Governing the Urban in China and India: Land Grabs, Slum Clearance, and the War on Air Pollution*. Princeton, NJ: Princeton University Press.

Schneider, Mindi. 2017. "Wasting the Rural: Meat, Manure, and the Politics of Agro-industrialization in Contemporary China." *Geoforum* 78: 89–97.

Schwenkel, Christina. 2013. "Post/Socialist Affect: Ruination and Reconstruction of the Nation in Urban Vietnam." *Cultural Anthropology* 28 (2): 252–77.

Scott, James C. 1998. *Seeing Like a State: How Certain Schemes to Improve the Human Condition Have Failed*. New Haven, CT: Yale University Press.

Shi, Jiachun, Gang Wang, Yan He, Jianjun Wu, and Jianming Xu. 2010. "Lead Accumulation in Westlake Longjing Tea: Non-Edaphic Genesis as Revealed by Regional Scale Estimate." *Journal of Soils and Sediments* 10: 933–42.

Siu, Helen. 2007. "Grounding Displacement: Uncivil Urban Spaces in South China." *American Ethnologist* 34 (2): 329–50.

Sivaramakrishnan, K., and Arun Agrawal, eds. 2003. *Regional Modernities: The Cultural Politics of Development in India*. Stanford, CA: Stanford University Press.

Solomon, Harris. 2016. *Metabolic Living: Food, Fat, and the Absorption of Illness in India*. Durham, NC: Duke University Press.

Stoler, Ann Laura, ed. 2013. *Imperial Debris: On Ruins and Ruination*. Durham, NC: Duke University Press.

Swislocki, Mark. 2009. "Culinary Nostalgia: Regional Food Culture and the Urban Experience in Shanghai." Stanford, CA: Stanford University Press.

Tsing, Anna Lowenhaupt. 2015. *The Mushroom at the End of the World: On the Possibility of Life in Capitalist Ruins*. Princeton, NJ: Princeton University Press.

UNESCO. 2018. World Heritage Center. *West Lake Cultural Landscape of Hangzhou*. Accessed January 8, 2018. http://whc.unesco.org/en/list/1334/.

Wang, Hongxin. 2009. *Land Policy Studies* [土地政策学]. Beijing: Beijing Normal University Press.

Wang, Li, Zhong Zhang, Yujie Fan, Yutong Xie, Sun Xiong, eds. 2014. *Penghu Qu, Chengzhong Cun, Feiqi Di Geng Xin de Shijian Tansuo yu Jingdian Anli* [Practical experiments and classic cases in the revitalization of shantytowns, urban villages, and wastelands]. Beijing: China Building Materials Press.

Wang, Liping. 1999. "Tourism and Spatial Change in Hangzhou, 1911–1927." In *Remaking the Chinese City: Modernity and National Identity, 1900–1950*, edited by Joseph W. Esherick, 107–21. Honolulu: University of Hawai'i Press.

Williams, Raymond. 1973. *The Country and the City*. New York: Oxford University Press.

Wu, Ying. 2019. "Becoming New Urbanites: Residents' Self-identification and Sense of Community in 'Village-Turned-Community.'" *The Journal of Chinese Sociology* 6: 7.

Xixi National Wetland Park. "Xixi Gaikuang" [General introduction to Xixi]. Accessed January 8, 2018. http://www.xixiwetland.com.cn/access_xixi.html.

Ye, Jingzhong. 2015. "Land Transfer and the Pursuit of Agricultural Modernization in China." *Journal of Agrarian Change* 15 (3): 314–37.

Yeh, Emily T. 2009. "From Wasteland to Wetland? Nature and Nation in China's Tibet." *Environmental History* 14: 103–37.

Zhao, Fang-Jie, Yibing Ma, Yong-Guan Zhu, Zhong Tang, and Steve P. McGrath. 2015. "Soil Contamination in China: Current Status and Mitigation Strategies." *Environmental Science & Technology* 49 (2): 750–59.

Contributors

Anne Rademacher is professor of environmental studies at New York University. Her work explores urban ecology and the future of cities. Her most recent book is *Building Green: Architects and the Struggle for Sustainability in Mumbai* (2017, University of California Press).

K. Sivaramakrishnan is Dinakar Singh Professor of Anthropology, professor in the School of the Environment, co-director of the Program in Agrarian Studies and co-director of the Inter Asia Initiative at Yale University. His work ranges across environmental history and ethics, rural development and wildlife conservation, migration and rainfed agriculture, and environmental law and jurisprudence in India. His most recent book, co-edited with conservation biologist Ghazala Shahabuddin, is *Nature Conservation in the New Economy: People, Wildlife and the Law in India* (2019, Orient Blackswan).

Andrew Alan Johnson is a cultural anthropologist who specializes in environmental and urban change in Southeast Asia. He received his PhD from Cornell University in 2010 and has taught at Princeton University, Yale-NUS College, Columbia University, and other institutions. He is the author of *Ghosts of the New City* (2014, University of Hawai'i Press), *Mekong Dreaming* (2020, Duke University Press), and numerous journal articles. He is currently a visiting scholar at the Center for Southeast Asian Studies, University of California, Berkeley.

Camille Frazier is assistant professor of anthropology in the Department of Humanities and Social Sciences at Clarkson University. Her research examines the relationships among shifting food systems and urban ecologies in order to understand sociopolitical and economic anxieties related to urbanization and environmental change. Her current book project considers these themes in the context of Bengaluru, India.

Kasia Paprocki is an assistant professor in the Department of Geography and Environment at the London School of Economics and Political Science. Her work draws on and contributes to the study of political ecology and the political economy of development and agrarian change.

Annu Jalais is an environmental anthropologist. She works as an assistant professor at the National University of Singapore's South Asian Studies and Comparative Asian Studies Departments. Her interdisciplinary research and teaching experience focuses on the human-nonhuman interface, environment and climate change, religious identity and migration, caste, and social justice. Her primary region of specialization is South Asia, specifically Bangladesh and India, and her secondary zone of interest encompasses Southeast Asia and China, especially around Indian Ocean exchanges in the religious and cultural realms.

Harris Solomon is associate professor of cultural anthropology and global health at Duke University. His interests bridge medical anthropology, science and technology studies, South Asian studies, and public health. He is the author of *Metabolic Living: Food, Fat, and the Absorption of Illness in India* (2016, Duke University Press) and *Lifelines: The Traffic of Trauma* (forthcoming, Duke University Press).

Tomonori Sugimoto received his PhD in anthropology from Stanford University. He was previously a postdoctoral associate in the environmental humanities at Yale University.

Shubhra Gururani is the chair and associate professor in the Department of Anthropology, York University. Her research focuses on the politics of land, planning, and ecologies in the cities of the Global South.

Erik Harms is associate professor of anthropology and chair of the Council on Southeast Asian Studies at Yale University. He is an urban anthropologist with long-term research interests in the social and spatial transformations of Saigon–Ho Chi Minh City. He is the past president of the Association for Political and Legal Anthropology, and the author of *Luxury and Rubble: Civility and Dispossession in the New Saigon* (2016, University of California Press) and *Saigon's Edge: On the Margins of Ho Chi Minh City* (2011, University of Minnesota Press).

Caroline Merrifield received her PhD from Yale University in cultural anthropology. Her research investigates food sourcing arrangements in China's alternative food movement, which has emerged against a background of urgent public concern over food safety. Caroline's work engages with theories of kinship, trust and risk, agrarian change and urbanization, taste and flavor, and the state. She is currently preparing a book manuscript based on her dissertation project, titled *Flavor and Fellowship: Cultivating Food System Alternatives in China*.

Index

Note: Page numbers set in italics refer to illustrative material (figures and tables).

accidents: fire at Santika club, 35–37; intentionality vs. unintentionality of, 31–32; motorcycle, 107, 108, 112, 113–14; shrines as, 33; spirits and legacy of, 27; transformative power of, 39, 40–41. See also bodily injuries
accountability, 106, 107, 110
activism: A. Agarwal on social equity and environmental, 5; about Mumbai's potholes, 105–6, 112–15; Baviskar's critique of, 11; by farmers in the Sundarban region, 66–67, 70; class exclusion and environmental, 59; in Taipei, 122, 131, 132, 135
A Dai, 177, 181–82, 183, 184, 186, 189
adaptation. See climate change adaptation (Sundarban region)
Agarwal, Anil, 5
Agarwal, Chetan, 145
agency, 31–32, 33, 36, 41
agraharam (residential enclave), 12
Agrawal, Arun, 179, 182, 187
agriculture. See farming
"agri-tainment" farms, 87–90, 122
air conditioning, 167–68
airports, 30
A-kuan (Sawmah's husband), 133, 134
'Amis peoples. See Pangcah/'Amis peoples
Amrith, Sunil, 18
Amrute, Sareeta, 108
Anand (gardening enthusiast), 49, 51, 52, 54, 57, 58
Anand, Nikhil, 104, 164

animals: consumption of wild, 91–92; displaced, 7–9, 30; on Singaporean "agri-tainment" farms, 89; statues of, 26–27, 33; use of term, 85n7. See also nonhuman life and environment
animism, 33
Anthropocene, 1–2, 4, 77–78, 96–97
anthropology, 33, 122
apps, 111
aquaculture, 66–67, 69–73, 74, 76
architects. See urban designers, developers, and planners
art, 109–10, 111, *112 fig. 5.2*, 154
Ashley, Jennifer, 110–11
Asian cities, approach to, 2, 5, 6–11, 12–20, 83. See also names of individual
Asif (victim of pothole), 107–8
assetization, 142, 147–51
automobiles, 168
"Away from the Devil and the Deep Blue Sea" (WWF), 75–76
Azande society, 31–32

Badshahpur drain, 138, 141
Baldy (Grange employee), 178
banana tree spirits, 26–27, 29, 34
Banay (Kilang resident), 132
Banerjee, Mamata, 73
Bangalore. See Bengaluru (formerly Bangalore)
Bangkok, 16, 160, 164. See also tree spirits (Bangkok)
Bangkok Metropolitan Authority (BMA), 26, 28, 33

Bangladesh, 68, 70, 72–74, 163. *See also* climate change adaptation (Sundarban region)
Bangladeshi migrants. *See* migrants and migration: Bangladeshi
Baviskar, Amita, 11, 121–22, 123, 131
Bear, Laura, 142
"Beasts, People, Wild Environments" (university course), 85
belonging, 74, 127, 135
Belton, Ben, 71n9
Bengaluru (formerly Bangalore), 12, 13, 18, 111, 140. *See also* organic terrace gardens (Bengaluru)
Bennett, Jane, 31, 33
Bhattacharya, Neeladri, 150–51
Bhattacharyya, Debjani, 7, 67, 142, 149–50
Bhumibol Adulyadej, King, 28
bicycle-share programs, 27, 28
Big River in Front of My House, A (film), 130
Bình Xuyên, 165n5
biophysical processes, approach to, 1, 5, 6, 10
Bipin (victim of pothole), 112–15
birds, 187–88
Björkman, Lisa, 164
BJP (political party, India), 52, 73
BMA (Bangkok Metropolitan Association), 26, 28, 33
BMC (Bombay Municipal Corporation), 105, 109, 113–14
bodily injuries: approach to, 104; Asif's, 107–8; Bipin's, 112, 113, 114; deaths from potholes, 102–3, 108; show multiple ecologies, 115; statistics on, 105, 106. *See also* accidents
Bollywood Veggies (farm), 89, 90
Bombay. *See* Mumbai
Bombay High Court, 106, 114n8
boundaries: community-environment, 127; land-water, 140, 142, 144, 146–47, 150
Brahmins, 12n18
Britain, 7
British East India Company, 67
British rule. *See* colonialism
Brook, Barry W., 93

Buddhism, 30, 34, 39–41
Building and Construction Authority (Singapore), 82n3
building orientation, 171
built *vs.* natural environment, 126–27
Bukit Brown Cemetery, 84

café society, 168
Calcutta. *See* Kolkata (formerly Calcutta)
capitalism, 69, 180–81
Carson, Rachel, 3
castes, 12n18, 48, 56–60, 143, 148, 151. *See also* class; inequalities; middle class
Catholic church, 134
cats, 92, 94–96
cemeteries, 36, 84
Center for American Progress, 72
Centre for Science and Environment, 139
certification of organic foods, 47, 53
Chakrabarty, Dipesh, 1
Chakravarty-Kaul, Minoti, 151
Chan, Ying-kit, 88
Chao Phraya River, 26, 27
Charlop-Powers, Sarah, 13
Chee, Lilian, 94–95
Chen Dong-sheng, 128–29
Chen Jiuhe, 188
Chennai, 12–13
Chennappa (leader in VK community), 48, 57, 58–59
Chiang Kai-shek, 124
Chiang Mai, 34
chickens, 91
children, 55
Children in Heaven (documentary), 130
China, 91, 92. *See also* foodshed (Hangzhou)
Chinese, Han, 120n1, 123n8, 182n6
Chinese female migrants, 95
Choudhry, Chetna, 138
Chowdhury, Prem, 139n2
Chu, Julie, 105
Chuang, Julia, 182
chuncai (plant), 183–85, 190
cities, fluidity of, 5–6, 10n14
city and urban spirits, 30, 33–34

City Improvement Trust Board (CITB, Bengaluru), 48
city-making. *See* urban development and planning
Clancey, Gregory, 97
class, 59–61, 108, 114, 165. *See also* castes; inequalities; middle class
cleanup of cities, 8–9
climate change. *See* sea level rise
climate change adaptation (Sundarban region), 64–81; alternative visions of, 68–69, 77, 78; approach to, 17–18, 64n2, 65–66, 77; as opportunity for new moral ecologies, 67–68; migration to New Town, 73–74; Saigon and, 163; shrimp cultivation in Kolanihat, 66–67, 69–73, 74, 76; use of term, 66; WWF's vision for, 64–65, 75–76
climate migrations, 71–72
Cobra Swamp, 30
Cold Storage (supermarket), 91
Collard, Rosemary-Claire, 69
colonialism: botanical gardens as project of, 18; *huang* and, 191; in India, 48, 56–57, 67, 138–39, 149–51; in Taipei, 124; in Thailand, 34; in Vietnam, 159
"Coming in to the Foodshed" (Kloppenburg et al.), 179–81
commons, 148, 150–51. *See also* urban ecological commons (Taipei region)
communities: housing, 128–29, 150, 153–55, 167 (*See also* public housing); in Gurgaon, 144–45; in Sandy Banks, 186; multigenerational co-habitation, 168; of food producers and consumers, 180. See also *niyaros* (community/home)
concrete, 30, 162, 166, 171–72
Congress party (India), 73, 148n9
connectivity and holes, 106
conservation, 5, 13, 86–87. *See also* green projects and spaces
cosmopolitanism, 27–28
COVID-19, 90, 91
cows, 8
Cronon, William, 191

crosses, 36
crows, 92–94, 95–96
Cruikshank, Julie, 32
CS (Ghata resident), 150, 151
culling, 85n6, 92–96
cultivation, 66–67, 69–73, 74, 76. *See also* foodshed (Hangzhou); fruit and vegetable cultivation; gardens
cultural identity, Singaporean, 84, 86, 90, 92
culture. *See* nature/culture interface; social processes

da Cunha, Dilip, 144
"Dada" tree, 154
Dahan River, 124, 126, 128, 130
dances, 36, 37 *fig. 1.2*
Das, Veena, 140
Davis, Lucy, 94–96
death of nature: approach to, 1–25, 83; connotations of term, 2–3
Death of Nature, The (Merchant), 3
De Boeck, Filip, 109
debt, 72
Deccan Herald, 44
DeCerteau, Michel, 30–31, 33
deities. *See* tree spirits
Delhi, 8–9, 138–39, 142
Delhi-Gurgaon expressway, 138
deltas, 164
Department of Town and Country (DTCP, India), 149
depeasantization. *See* displacement and dispossession; migrants and migration
designers. *See* urban designers, developers, and planners
DeSo architectes, 172–73
Detailed Master Plan and Urban Design Guidelines for Thủ Thiêm New Urban Area, The, 170–71
developers. *See* urban designers, developers, and planners
development. *See* urban development and planning
Dhaka, 72, 74, 163
digging into numbers, 38, 39 *fig. 1.3*
"dirty" food, 91–92, 132

displacement and dispossession: approach to, 15; as climate change adaptation strategy, 75; in Delhi, 9; in Hangzhou, 178–79, 182, 184, 186, 187, 188–89; in Saigon, 165–66, 172, 173–74; in Taipei, 123, 127, 129–30, 131, 132, 135; in the Sundarban region, 65–66, 69–73, 74; of animals, 7–9, 30; of material absence and presence, 102; of OTGians, 54n7; of VK community, 48. *See also* migrants and migration
District Gazetteer of Gurgaon, 141
Dixon, Terrell, 14
DLF (land developer), 139, 148, 150
domestic space, 94–95
dongec (rattan's stem), 133, *134 fig. 6.3*
Donovan, Deanna G., 92
Douglass, Mike, 163–64
draft plans, 148–49
drains, 138, 139n3, 141
Duy, Phan N., 163
dwelling (as term), 126–27, 185n7

ecological consciousness: of Saigon urban designers, 170–73; of Thủ Thiêm residents, 167–69
ecological design, 73, 161, 170–73. *See also* green projects and spaces
Ecologies and Urbanism in Asia III (workshop), 85–86
ecologies of urbanism: approach to, 1–25, 77, 83, 123; use of term, 10n14, 12, 65, 83
Ecologies of Urbanism in India (Rademacher and Sivaramakrishnan), 10, 83–84
ecosystems: approach to, 11–12; in Saigon, 161, 169, 173; organic terrace gardeners' desire to build, 54; Sundarbans as unique, 67; water management and, 164
Edible Garden City (farm), 89
edible gardens. *See* organic terrace gardens (Bengaluru)
e-governance, 113, 114
Elinoff, Eli, 30
Elvin, Mark, 182n5

Empire and Ecology in the Bengal Delta (Bhattacharyya), 149–50
End of Nature, The (McKibben), 3
endosulfan poisoning, 51
environment. *See* nonhuman life and environment
environmental humanities, 4–5, 12
Environmental Humanities (journal), 4
environmentalism, 27–28. *See also* activism
environmental justice, 2
environmental protection, 5, 13, 86–87. *See also* green projects and spaces
environmental threats, 152. *See also* sea level rise
equity, 5. *See also* inequalities
Escobar, Arturo, 123, 127
Europe and European worldviews, 1–2, 7, 8, 27, 34. *See also* colonialism; Western worldviews
Evans-Pritchard, Edward Evan, 31–32
exclusion: approach to, 15; environmental activism and class, 59; environmental concerns used to justify, 164; in organic terrace gardening, 48; of migrants in New Town, 74; of nonhuman life, 2; of OTGians, 54n7. *See also* inequalities
extinction, 3

Facebook, 45, 49, 88, 110, 113
fairs, 49–51
farmers. *See* food producers
farming: for "agri-tainment," 87–90, 122; industrial, 180; in green urbanism, 168; in Xixi Wetland, 188, 189–90. *See also* organic terrace gardens (Bengaluru); urban ecological commons (Taipei region)
farming futures and livelihoods: in Gurgaon, 144, 153; in Hangzhou, 179, 182, 184–85; in the Sundarban region, 64–67, 69–73, 74–76, 77–78; obscured, 142–43. *See also* food producers
farmlands: as *huang*, 190–92; and urban development in Bengaluru, 57; and urban development in Gurgaon, 139,

147–50, 151, 154; and urban development in Hangzhou, 179, 181–82, 184, 185–87, 190
fauna. *See* animals; nonhuman life and environment
female spirits *(nang mai)*, 26–27, 29–30, 32, 34–40
Ferguson, Jane, 30
Fernandes, Leela, 46
Fill in the Potholes Project, 111
fire, 35–37
fishing, 91–92, 125–26, 131–32, 169, 188
flooding: in Bangkok, 26, 160, 164; in Gurgaon, 138, 139–40, 155–56; in Mumbai, 102, 163. *See also* monsoons
flood talk in Saigon: approach to, 159–60, 162; by residents, 160–61, 164–67, 173; by scientists, 162–64
flora. *See* nonhuman life and environment; plants
food: eaten by Pangcah/'Amis peoples, 120, 126, 132, 133–35; from Xixi Wetland, 189; Singaporean sources of, 89, 90–92. *See also* fruit and vegetable cultivation; organic terrace gardens (Bengaluru)
food insecurity, 49
food producers: in Bengaluru, 45–46, 48, 52–53, 58–60; in Singapore, 91; Sundarban region farmers, 65–67, 69–73, 74; supplying the Grange, 177–78, 180, 182–85, 187. *See also* farming futures and livelihoods
food retailers, 53, 91
food safety, 44–45, 49, 51–53, 58, 91–92, 177
foodshed (Hangzhou), 177–95; approach to, 19–20, 181, 182–83, 190; *chuncai* in, 183–85, 190; foodsheds in India and, 45; *huang* land and, 190–92; procurement team at the Grange, 177–78; Sandy Banks Village and, 185–87, 190, 192; use of term, 179–80; Xixi Wetland and, 178, 187–90
foraging, 91, 132, 169
foreigners, 38–39, 82n2. *See also* migrants and migration

foreign species, 93–94
forests, 64, 82. *See also* tree spirits (Bangkok)
French rule, 159
fringe foods, 47
fruit and vegetable cultivation: in Bengaluru, 47, 48, 58; in Taipei, 120, 132, 133–35; on Singaporean "agritainment" farms, 88
Funahashi, Daena, 34
futures. *See* farming futures and livelihoods; urban futures and livelihoods

Gandy, Matthew, 191
Ganges River, 6–7
Gang of Sanchong (developer), 129
garbage collection. *See* waste management
Gardenasia (farm), 88–89
Garden City: Bengaluru as, 54, 57; Singapore as, 82, 86
gardens: in Hangzhou, 192; in Taipei, 120, *121 fig. 6.1*, 122, 132, 133–35; photo of VK, *60 fig. 2.4*; private *vs.* public, 18. *See also* organic terrace gardens (Bengaluru)
Garrard, Greg, 4, 5
Garschagen, Mattias, 164
Ge, Master, 183–85, 187
gender, 95–96, 108
genetically modified organisms (GMOs), 52
Ghata Jheel (Ghata Lake): locals on, 141–42, 150; mapping, 143–44; no steps taken to restore, 153; paved over, 150, 151; squatting by migrants in, *152 fig. 7.1*
Ghata village, 141, 144, 150, 151
ghers (shrimp cultivation ponds), 70–71
Ghosh, Amitav, 3
Ghosh, Nilanjan, 76
ghosts, 95. *See also* tree spirits (Bangkok)
Gidwani, Vinay, 121–22, 123, 131
Goh, Daniel, 84–85, 87
Goldman, Michael, 140
government, Haryana, 142, 145–56, 148–49, 152–53
government, Indian, 52, 106

government, Mumbai, 102, 105, 106, 109–10, 113–14
government, Punjabi, 139
government, Singaporean: citizens kowtowing to *vs.* rebelling against, 87, 96, 97; land acquisition by, 84, 88; notions of nature, 84–85; policies on crows and cats, 92, 94–96; promotion of tourism by, 82–83, 87, 90
governments, Chinese, 184, 185, 186, 188, 191
governments, Taiwanese: demolish *niyaros*, 129–30; inability to eliminate urban ecological commons, 123, 130; KMT Party, 124–25, 127; Pangcah/'Amis peoples challenge, 130, 131, 135–36; policies for "cleaning up" riverbanks, 128, 131
graduate mother scheme, 96
grains, 47
Grand Arch (housing community), 150, 153–55
Grange (organic restaurant): approach to, 181, 182–83; *chuncai* at, 183–85, 190; current procurement area, 179; *huang* and, 191–92; and Kloppenburg's vision of foodshed, 180, 181; origins and philosophy, 177; procurement team, 177–78; senior employees, 186; Xixi Wetland as former procurement area for, 189
graveyards, 36, 84
green architects, 170–73
green projects and spaces: approach to, 14–15, 18; as motivation for organic terrace gardening, 53–56, 60; as post-industrial ecologies of nature, 13; in Bangkok, 27–28, 32; in Gurgaon, 153; in Hangzhou, 178, 187–90; in Taipei, 122–23, 125, 127, 128–30, 131, 135–36; motivation for Asian cities', 10; types of parks in Asian cities, 18. *See also* Singapore; sustainable urban development
Green Revolution, 49, 51, 52
green urbanism, manifesto of, 168, 170

greenwashing, 172
grievances, 112–15
grocery stores, 91
gross domestic product (GDP), 82n1
Gujjars, 139, 141–42, 144–45, 155
Gupta, Akhil, 115
Gupta, Subodh, 154
Gupte, Rupali, 111, *112 fig. 5.2*
Gurgaon (now Gurugram), 138–58; approach to, 17, 140, 155–56; assetizing land and water in, 142, 147–51; and defining water bodies, 143–47; flooding in, 138, 139–40, 155–56; Grand Arch housing community in, 150, 153–55; movement to restore water bodies in, 152–53
Gurgaon District, 140–41
Gurgaon residents, 139, 141–42, 144–45, 150, 151, 155
Gurugram Metropolitan Development Authority (GMDA), 142, 145, 153
Gurujam, 138
Gu Yingchun, 185

Han Chinese, 120n1, 123n8, 182n6
Hangzhou. *See* foodshed (Hangzhou)
Hangzhou World Cultural Heritage Monitoring and Management Center, 186n13
Haraway, Donna, 65, 68–69, 77
Harms, Erik, 58, 67, 135, 191n18
Harvey, David, 14n20
Haryana: government in, 142, 145–46, 148–49, 152–53; in National Capital Region, 147; Punjab and, 139, 150
Haryana Pond and Waste Water Management Authority, 145–46, 152–53
Haryana Pond and Waste Water Management Authority Act (India), 152
Haryana State Pollution Control Board (HSPCB), 153
Hay, Leon, 89
Hay Goat Dairies (farm), 89

Index

HDB (Housing and Development Board, Singapore), 88n10, 94
health, 90, 91, 93, 94, 96, 97. *See also* food safety
heat gain, 171
Hedden, W. P., 179
Heidegger, Martin, 126
Heller, Patrick, 46
Hendrickson, John, 179–81
Heng, Geraldine, 96n17
Herzfeld, Michael, 16, 28
hillsides in Taipei: abundance, 124, 125; Icep, 120–21, 126, 132–35; urban development on, 123, 127, 128–30
Hindu, The, 50–51
Hinduism, 12, 29, 30
historical context, approach to, 6–9
Ho Chi Minh City. *See* Saigon
Hooda, Bhupinder Singh, 148–49
hospitals, 103, 107, 108, 115
Housing and Development Board (HDB, Singapore), 88n10, 94. *See also* public housing: in Singapore
housing communities, 128–29, 150, 153–55, 167. *See also* public housing
How Great Cities Are Fed (Hedden), 179
Huang, Mister, 178
huang (abandoned or fallow) land, 190–92
Hybrid Geographies (Whatmore), 97

Icep (community), 120–21, 126, 130, 132–35
Ichang (Kilang resident), 130–31
identity, Singaporean cultural, 84, 86, 90, 92
imaginaries and imagination, 4, 27, 135, 140, 142–43. *See also* climate change adaptation (Sundarban region)
Imperial Debris (Stoler), 181
impermeability and permeability, 102, 106, 162, 166
India, 7, 17, 68, 72–74, 91. *See also* climate change adaptation (Sundarban region); Gurgaon (now Gurugram); organic terrace gardens (Bengaluru); potholes (Mumbai)
Indian Institute of Science, 44

Indian migrants, 68, 88, 90
indigenous peoples, 124, 128. *See also* Pangcah/'Amis peoples
Industrial and Information Technology Training Institutes, 76
industrialization, 6–8. *See also* urban development and planning
inequalities: Amrith on climate change and, 18; approach to, 2, 13; Baviskar on, 11; in Singapore, 88, 90, 97; organic terrace gardening and, 46, 57, 59; potholes and, 104. *See also* castes; class; displacement and dispossession; exclusion; marginalized peoples
information technology (IT) professionals, 49, 51, 55–56
infrastructures, 7, 26–27, 31, 33, 75–76, 138. *See also* potholes (Mumbai)
Ingold, Tim, 126, 185n7
injuries. *See* accidents; bodily injuries
Intergovernmental Panel on Climate Change, 6
introduced species, 93–94
invasive species, 93–94
IREO (land developer), 150, 153–54
irrigation systems, 138–39

Jain, Lochlann, 107
Jalais, Annu, 67
Jal Mitra (organization), 13
jao, 32n1. *See also* tree spirits (Bangkok)
Jao Hai Tek (spirit), 29
Jats (caste), 148, 151
Jiap (tree spirit devotee), 38–40, 41
johads. *See* ponds

Kaika, Maria, 13n20
kampungs, 88
Kapaleeswarar Temple, 12
Karnataka Department of Horticulture, 51, 52
KCA (Kranji Countryside Association), 87–90, 122
Keelung River, 124, 128
Kerala, 51
Ketagalan tribe, 124

Khan, Sameera, 108
Khulna City, 70, 72
Khulna District, 69, 72
Kilang (community), 125–26, 130–32
Kim, Eleana, 105
Kinshasa, 109
Kloppenburg, Jack, 179–81
KMT Party (Kuomintang), 124–25, 127
Kolanihat, shrimp cultivation in, 66–67, 69–73, 74, 76
Kolbert, Elizabeth, 3
Kolkata (formerly Calcutta), 72–74, 75–76, 149–50, 163
Kong, Lily, 90–91, 92
Kothari, Smitu, 123
Kranji Countryside Association (KCA), 87–90, 122
KS (Gujjar elder), 141–42, 155
Kufi (Kilang resident), 131–32

labor market, 38, 74, 125. *See also* farming futures and livelihoods
lady mothers (female spirits), 26–27, 29–30, 32, 34–40
lakes, 12, 13, 140, 187–88. *See also* Ghata Jheel (Ghata Lake)
Lamont, Mark, 106–7
land. *See* boundaries: land-water
land acquisition: in Bengaluru, 48; in Gurgaon, 147–49, 150; in Kolanihat, 70; in Singapore, 84, 88; in the Sundarban region, 75
Land Acquisition Act (India), 148–49
Land Acquisition Act (Singapore), 84, 88
land assetization, 142, 147–51
land classification. *See* mapping
land developers. *See* urban designers, developers, and planners
land elevation, 165, 166
Land Reform Act (India), 151
land scarcity, 84, 87
land tenure arrangements, 151, 186–87
language, 38n6, 46, 50–51, 144, 161–62, 173
latency, 135
Latour, Bruno, 3–4, 153
law, 105–6, 107

Lee Kuan Yew, 83, 86
Lefebrvre, Henri, 5
legumes, 47
Lek (tree spirit devotee), 40
Le Phan Cement Corporation, 171
Lévi-Strauss, Claude, 160
Lim, Haw Chuan, 93
Lim Kim Seng, 93
livestock, 8
London, 7
Longshanlin (housing community), 129
Lorimer, Jamie, 96–97
Low, Bjorn, 88–89

MacRitchie reservoir, 91–92
Malabou, Catherine, 40
male spirits, 29–30, 34
Manchester Guardian, 7
mangrove forests, 64
manholes. *See* potholes (Mumbai)
manifesto of green urbanism, 168, 170
Mao, Auntie, 183–85, 187
mapping, 141, 142, 143–44, 151, 165
Marathi songs, 109–10
marginalized peoples: approach to, 15; desire to transcend their current lives, 39, 40, 41; in Saigon, 165–66; rural poor, 34. *See also* inequalities; migrants and migration; Pangcah/'Amis peoples; urban poor
market-oriented gardening, 48, 56–59. *See also* food producers
Marks, Danny, 160
marshes. *See* wetlands and marshes
master plans, 139, 148–50, 170–71
Mathur, Anuradha, 144
Mayaw (Kilang resident), 125–26, 130, 132
Mayaw Biho, 130
Ma Zhihui, 189
Mazzarella, William, 114
McKibben, Bill, 3
media: and class in India, 114; on Bengaluru, 44; on flooding in Saigon, 166; on Gurujam, 138; on invasive species in Singapore, 94; on Mumbai's potholes, 105, 108, 113; on Oota From

Your Thota, 50–51; on Singaporean "agri-tainment" farmers, 88–89; on Thames River, 7; social media, 105, 138 (*See also* Facebook)
Mendonsa, Malishka, 109–10, 116
Merchant, Carolyn, 3
Merrifield, Caroline, 45
metro, 104
microcredit loans, 72
middle class: in Bangkok, 27; in Bengaluru (*See* organic terrace gardens (Bengaluru)); in Saigon, 165; in Taipei, 127–29; speculation bolsters aspirations of, 142; Sundarban region tourism by, 76; use of term, 46. *See also* castes; inequalities
migrants and migration: approach to, 15; Bangladeshi, 67, 68, 70, 71–74, 75–76, 88, 90; increase in, 5–6; in Singapore, 88, 89–90, 95; in Taipei, 129; Pangcah/'Amis peoples, 125; squatting in Ghata Jheel by, *152 fig. 7.1*; tree spirit devotees, 38. *See also* displacement and dispossession
milk, 8
Miller, Michelle, 163–64
Ming (Grange employee), 178
"Ministry of Potholes" (Twitter account), 105
Ministry of Road Transport and Highways (India), 106
Mintz, Sidney, 47
mixed-use zoning, 168
modernism, 27, 28, 30
Modi, Narendra, 52, 73
monarchy, 27–28, 30
monkeys, 9
monsoons, 102, 107–8, 109–10, 115, 116, 156. *See also* flooding
moral ecologies, 66–78; alternative, 77; climate change as opportunity for new, 67–68; defined, 66; in Saigon, 67, 165; Kolanihat and, 69–73; of imagined erasure, 75–76; of migration to New Town, 73–74; render farming futures obsolete, 78; Xixi Wetland and, 188

Morrison, Kathleen, 1–2
Moses, Robert, 31
mosquitoes, 95
Mother Dairy, 8
motorcycle accidents, 107, 108, 112, 113–14
multigenerational co-habitation, 168
Mumbai, 140, 164, 172. *See also* potholes (Mumbai)
Mumbaikars: app for, 111; Asif, 107–8; Bipin, 112–15; holes as somatic and, 106; "Ministry of Potholes" and, 105; pothole songs by, 109–10
Mumbai Metro, 104
Mumbai Metro Rail Corporation, 104
Mushroom at the End of the World, The (Tsing): climate change adaptation and, 65, 68, 77, 78; foodshed in Hangzhou and, 180–81, 184–85, 191; urban ecological commons and, 121–22, 123, 130, 135
Mustafa (grocery store), 91

Nagendra, Harini, 13
Najafgarh drain and Sahibi River, 139, 141
Nakaw (Icep resident), 134–35
nang mai (female spirits), 26–27, 29–30, 32, 34–40
Nappi, Carla, 92
Nasir (Asif's friend), 107–8
National Capital Region (NCR, India), 147, 148
National Forestry Administration (China), 187
National Green Tribunal (NGT), 145, 152
National University of Singapore, 85
native peoples, 124, 128. *See also* Pangcah/'Amis peoples
Natural Areas Conservancy, 13
natural *vs.* built environment, 126–27
nature/culture interface: inseparability of, 32; in Singapore, 84; Rademacher and Sivaramakrishnan on, 77, 83
nature spaces. *See* green projects and spaces
nature spirits. *See* tree spirits
Nature Watch, 93
Navalkar, P., 103

NCR (National Capital Region, India), 147, 148
negligence, 103, 108–9, 110–11, 113–14
neighbors and terrace gardening, 54
New Delhi, 138–39
New Gurgaon, 148
newspapers. *See* media
New Taipei City. *See* urban ecological commons (Taipei region)
New Town, 73–74
New Urban Zones (Saigon): as cause of flooding, 160; as reminder of city's unique ecology, 174; moral ecologies and, 165; Thủ Thiêm, *161 fig. 8.1*, 166–73
New York City, 31
Nicholls, Robert, 159n1
Nilsson, Per Ake, 89
niyaros (community/home): community-environment boundaries in, 127; defined, 125n10; demolished, 129–30, 131; Icep, 120–21, 126, 130, 132–35; Kilang, 125–26, 130–32
nonhuman life and environment: approach to, 9, 14, 15; as part of humanity, 4; Chinese conceptions of, 182; climate change adaptation and, 67, 68–69, 77; exclusion of, 2; in Singapore, 85–86, 92–96; in Xixi Wetland, 187–88; Lorimer on, 96–97; nonhuman material, 31–32, 33, 41; use of term, 85n7. *See also* animals; plants
North America, 8, 31, 180. *See also* Western worldviews
nostalgia, 87–88, 90
numbers, digging into, 38, *39 fig. 1.3*

Oberoi, S. S., 152
Ocean of Wetness (design platform), 144
offerings, 34, *35 fig. 1.1*, 36, *37 fig. 1.2*
Ogden, Peter, 72
Ontology of the Accident (Malabou), 40
Oota From Your Thota, 49–51
Operation Flood, 8
organic market, 52–53

organic restaurants. *See* Grange (organic restaurant)
organic terrace gardens (Bengaluru), 44–63; approach to, 19, 45–46; and differing caste views on gardening, 56–59; and exclusion of marginalized castes, 48; fairs promoting, 49–51; food safety as motivation for, 49, 51–53, 58; green spaces as motivation for, 53–56, 60; origins, 49; OTGians' hands-on approach to, 56; photos, *47 fig. 2.1, 55 fig. 2.3*; and potential class alliances, 59–61; *vs.* urban ecological commons in Taipei, 122
OTGians (defined), 45
Other Backward Classes (OBC), 48
ownership of city spaces, 33, 40

Padwe, Jonathan, 69
Palafang (Icep resident), 120–21, 133
panchayat (village council), 150, 151
Pangcah/'Amis peoples: challenge government, 130, 131, 135–36; dwelling and, 127; in Icep, 120–21, 126, 132–35; in Kilang, 125–26, 130–32; migration of, 125; use of term, 120n2; view of public lands, 123
parks. *See* green projects and spaces
Parks and Recreation Department (PRD, Singapore), 83
parody, 110–11, 112
patriarchy, 154
Pearson, Trais, 31
Peasant Question, The (tract), 159
People's Action Party (PAP, Singapore), 84
permeability and impermeability, 102, 106, 162, 166
pesticides, 49, 51–52, 58, 91
Peters, Christian J., 179
pets, 8, 94–95
Phadke, Shilpa, 108
Phra Siam Thevarat (spirit), 29, 34
Phú Mỹ Hưng (New Urban Zone), 166
Pieprz, Dennis, 170

Index 209

Places of Nature in Ecologies of Urbanism (Rademacher and Sivaramakrishnan), 10, 83
planners. *See* urban designers, developers, and planners
planning. *See* urban development and planning
plants: *chuncai*, 183–85, 190; eaten by Pangcah/'Amis peoples, 126, 132, 133, 135; in green urbanism, 168; in Thủ Thiêm, 169. *See also* fruit and vegetable cultivation; nonhuman life and environment
Podesta, John, 72
Poison Ivy (restaurant), 90
police, 103, 107, 113
political action. *See* activism
political economies of development. *See* climate change adaptation (Sundarban region)
pollution, 6–7, 127–28
ponds: defined, 145–46; in Ghata village, 144; model, 153, 154; for shrimp cultivation, 70–71
poor people, 34. *See also* urban poor
population statistics, 45, 82, 125
post-industrial ecologies of nature, 13
"Pothole City" (art installation), 111, *112 fig. 5.2*
Pothole Festival, 110
potholes (Mumbai), 102–19; approach to, 16–17, 102, 103–4, 115; apps for finding, 111; Asif's injury, 107–8; Bipin's grievance report, 112–15; other advocacy about, 105–6; Saigon and, 163; songs and art about, 109–10, 112; worship of, 110, *111 fig. 5.1*, 112
power: and configurations of lives and livelihoods, 77; in Singapore, 95; Rademacher on ecologies and, 140; shaping moral ecologies, 66, 68–69, 71
Practice of Everyday Life, The (DeCerteau), 30–31
precolonial gardening, 18, 56
present-absences, 102, 104, 115
private lands, 151

private spaces, 46, 56–57, 122
procurement. *See* Grange (organic restaurant)
progress, 68, 69, 75–76, 78
property, 149–51
public housing: in Singapore, 88n10, 91, 92, 94, 96; in Taipei, 132
Public Interest Litigation (PIL), 105
public lands. *See* urban ecological commons (Taipei region)
pujas (worship), 110, *111 fig. 5.1*, 112
Punjab, 138–39, 150
Punjab Village Common Land Act (India), 151

Qian, Zhu, 179, 186n12
Qiu Qi-qian, 126n12

Rademacher, Anne: on anthropocentric hubris, 156; on belonging, 127; on cities' transformation of own urban ecologies, 163; on displacement, 123; on ecologies and power, 140; on environmental change-urban transformation interface, 83–84; on exclusion and water management, 164; on foregrounding of certain imaginaries, 143; on green architects, 172; on "making the environment," 32; on moral ecology, 66, 188; on multiple meanings of "urban nature," 59–60; on urban nature-urban culture interface, 77, 83; workshop held by, 85–86
radio stations, 109–10, 116
Raffles, Hugh, 32
Ragbar (Gujjar man), 139, 144–45
rains. *See* flooding; monsoons
Raj, 67
Rajinappa (young VK man), 60
Ramachandra, T. V., 44
Ranade, Shilpa, 108
Ranganathan, Malini, 146
Rattan's stem *(dongec)*, 133, *134 fig. 6.3*
real estate development: in Gurgaon, 139, 147–48, 149–50, 151, 153–55; in Taipei, 128–29

Redfield, Peter, 105n2
Red FM (radio station), 109–10, 116
region (use of term), 179
Regional Modernities (Sivaramakrishnan and Agrawal), 179
relocation. *See* displacement and dispossession
Report on Waterbodies of Gurugram (GMDA), 153
restaurants. *See* Grange (organic restaurant)
retreat, 75–76
Revenue Record, 145
Rhine Project, 131
rice farming, 66–67, 69, 71
rights. *See* displacement and dispossession; inequalities
riverbanks in Taipei: abundance, 125; approach to, 121; Kilang, 130–32; urban development on, 123, 127–30, 131
rivers: Chao Phraya, 26, 27; Dahan, 124, 126, 128, 130; Ganges, 6–7; Keelung, 124, 128; Sahibi, 139, 143; Saigon, 160, 165, 166; Tamsui, 124; Thames, 7; Xindian, 124, 126, 128, 129, 130–32
roads, 26–27, 33, 138. *See also* potholes (Mumbai)
Rocheleau, Dianne, 123
Rose, Deborah Bird, 4
Roy, Ananya, 67, 73
RS (real estate agent), 147, 149
ruins: climate change adaptation and, 65, 77, 78; Hangzhou foodshed and, 180–81, 182–83, 184–85, 190, 191
rural futures and livelihoods. *See* farming futures and livelihoods
rural people. *See* displacement and dispossession; farming futures and livelihoods
rural poor, 34
rural romanticised, 88, 89n12

Sahibi River and Najafgarh drain, 139, 143
Saigon, 159–76; approach to, 16, 159–60, 162, 173; Bengaluru and, 58; ecological consciousness of Thủ Thiêm residents, 167–69; land-water boundaries in, 140; moral ecologies in, 67, 165; residents on causes of floods, 160–61, 164–67, 173; scientists on causes of floods, 162–64, 166; urban designers' and planners' vision of, 160–62, 166–67, 170–73
Saigon RDC, 172
Saigon residents: ecological consciousness of Thủ Thiêm residents, 167–69; on causes of floods, 160–61, 164–67, 173
Saigon River, 160, 165, 166
sand, 160, *161 fig. 8.1*, 166
Sandy Banks Village, 185–87, 190, 192
Santika Club fire, 35–37
San'ying (community), 130, 131
Saowac (community), 130, 131
Sardar, Karunamoyee, 70
SARS, 91, 96
Sasaki Urban Design Associates, 170–73
Sawmah (Icep resident), 133–35
science and scientists: Anthropocene in physical, 4; language of, 161–62, 173; of the concrete, 160, 173; on causes of floods in Saigon, 162–64, 166
Scott, James C., 180
sea level rise: in Gurgaon, 140; in Saigon, 162–63, 164, 166; in Singapore, 97; and movement of cities, 6; Vietnam's vulnerability to, 159n1
sexuality, 108
shading devices, 167–68, 171
share systems, 186
Sheng Siong (grocery store), 91
shrimp aquaculture, 66–67, 69–73, 74, 76
shrines, 26, 27, 29, 33, 40
Shruthi (gardening enthusiast), 55–56
Siegel, James, 29
Silent Spring (Carson), 3
Simone, AbdouMaliq, 103
Singapore, 82–101; "agri-tainment" farms in, 87–90, 122; approach to, 19, 83–87; as model of green space, 27; culling of nonhuman life in, 92–96; ethnic composition of, 85; food sources in, 89, 90–92; Lorimer's ideas

Index

of Anthropocene applied to, 96–97; primeval vegetation type in, 82
Singaporeans: Chinese migrants and, 95; cultural identity, 84, 86, 90, 92; green spaces designed to lift spirits of, 83; kowtow to *vs.* rebel against the government, 87, 96, 97; notion of nature, 84–85; on culling of nonhuman life, 92–93; on food sources, 91–92; views on "agri-tainment" farms, 90
Singapore Tourist Promotion Board (STPB), 82–83, 87
Singh, Shalini, 149
Singh-Lim, Ivy, 89
Sinha, Neha, 17
Sinha, Vineeta, 90–91
Siu, Helen, 186n9
Sivaramakrishnan, K.: on anthropocentric hubris, 156; on belonging, 127; on cities' transformation of own urban ecologies, 163; on displacement, 123; on environmental change-urban transformation interface, 83–84; on humans as part of nature, 182; on the "processual," 187; on the term "region," 179; on urban nature-urban culture interface, 77, 83; workshop held by, 85–86
Sixth Extinction, The (Kolbert), 3
slow food, 168
slums, 9, 57, 73
snakes, 30
social justice, 143. *See also* inequalities
social media, 105, 138. *See also* Facebook
social processes, approach to, 1, 2, 5, 6, 10, 15, 20. *See also* inequalities
Sodhi, Navjot S., 93
Soh, Malcolm C. K., 93
Solomon, Harris, 52
songs, 109–10, 112
space: domestic, 94–95; ownership of city, 33, 40; private, 46, 56–57, 122; quest for, 3–4. *See also* green projects and spaces
speculation, 142
speculative urbanism, 140

spirits, 95. *See also* tree spirits
spiritual life, 168
Spothole (app), 111
squatting, *152 fig. 7.1*
Sra (Icep's founder), 126, 133, *134 fig. 6.3*
Srinath (gardening enthusiast), 54–56, 60
Srinivas, Smriti, 56
Statista, 82n2
statues, 26–27, 33
Staying with the Trouble (Haraway), 65, 68–69, 77
Stevenson, G. W., 179–81
Stevenson, Lisa, 105n2
Stoler, Ann, 181, 191
Storch, Harry, 163
Straits Times, The, 88–89, 94
students, 85, 90, 91–92
subsidence, 162, 166
Suenari, Michio, 125n10
Sundarban region. *See* climate change adaptation (Sundarban region)
supermarkets, 91
supernatural beings, 95. *See also* tree spirits
Survey of India, *145 table 7.1*
sustainability, 11, 143, 153, 177. *See also* green projects and spaces
sustainable urban development, 73, 161, 170–73. *See also* green projects and spaces
Sustenir Agriculture (farm), 88
Suvarnabhumi airport, 30

Taipei. *See* urban ecological commons (Taipei region)
tamarind trees, 29, 34–40
Tambiah, Stanley, 33
Tamsui River, 124
Tan, Audrey, 89
Tan Wei Xian, Alvin, 88–89
technology: in WWF's vision for Sundarban region, 76; IT professionals, 49, 51, 55–56; for locating potholes, 111, 113, 114; on Singaporean "agri-tainment" farms, 89; spirit answering prayers about, 29
temples, 12–13, 26

tenure arrangements, 151l, 186–87
terminology. *See* language
terrace gardening (defined), 47n7, 52.
    See also organic terrace gardens
    (Bengaluru)
Thames River, 7
Thủ Thiêm New Urban Zone, *161 fig. 8.1*,
    166–73
tigers, 67
Times of India, The, 51, 108
tourism, 76, 82–83, 87–90, 122, 187–88, 189
tradition in Hangzhou foodshed, 177, 179,
    180, 181
traffic, 54, 104–5, 138
tree of metal, 154
tree spirit devotees, 38–40, 41
tree spirits (Bangkok), 26–43; approach
    to, 15–16, 28, 32, 33; favors given by
    and sought from, 27, 29, 30, *35 fig.
    1.1*, 36–37, 38–40, 41; in banana trees,
    26–27, 29, 34; individuality of cults
    surrounding, 37; on site of nightclub
    fire, 34–40; shifting nature of, 28–29
Tsing, Anna. See *Mushroom at the End of
    the World, The* (Tsing)
Twitter, 105

ubatihet (a "happening"), 31
Uncle (A Dai's), 177–79, 185–86, 189, 192
UNESCO World Heritage sites, 187n15
United States, 8, 31, 180
university courses, 85
urban and city spirits, 30, 33–34
"Urban Commons" (Gidwani and
    Baviskar), 121–22, 123, 131
urban designers, developers, and planners: divine power harnessed by, 34;
    in India, 53, 139, 148, 149–50, 151,
    153–54; in Saigon, 160–62, 166–67,
    170–73; in Taipei, 129
urban development and planning: approach
    to, 6–9, 14–15; in Bangkok, 30, 33, 34,
    40; in Bengaluru, 45, 48, 54, 57, 59;
    in Gurgaon, 139, 142, 143, 146–47,
    148–50; in Hangzhou, 178–79, 181,
    182, 184, 185–87, 190, 191–92; in

Kolkata, 74, 75; in Mumbai, 104; in
    New York City, 31; in Saigon, 162–63
    (*See also* New Urban Zones (Saigon));
    in Singapore, 84, 88; in Taipei, 122–23,
    124–25, 127–30; in the Sundarban
    region, 66, 67–68, 69; Miller and
    Douglass on flooding and, 163–64;
    Ramachandra's challenge to narratives
    of, 44; reconciling social justice and
    sustainability with, 143; sustainable,
    73, 161, 170–73. *See also* green projects
    and spaces; real estate development
urban ecological commons (Taipei
    region), 120–37; abundance, 124–27;
    approach to, 16, 121–22, 123–24;
    decline, 127–30; Icep, 120–21, 126,
    130, 132–35; Kilang, 125–26, 130–32;
    Pangcah/'Amis peoples challenge
    government, 130, 131, 135–36; use of
    term, 121
urban food production. *See* organic terrace
    gardens (Bengaluru)
urban futures and livelihoods, 65, 73–75,
    76, 78, 185, 188–89
urban nature, approach to, 1–25, 77, 83, 123
urban poor, 2, 9, 15, 92, 97, 165
Urban Redevelopment Authority (URA,
    Singapore), 84, 87
urban villages, 186

Vahnikula Kshatriya (VK), 48, 57, 58–60
van Dooren, Thom, 93–94
Vaughn, Victoria, 94
vegetables. *See* fruit and vegetable
    cultivation
Venkatraman, T., 104
ventilation, 171
vibrant matter, 31–32, 33, 41
Vibrant Matter (Bennett), 31
Vietnam. *See* Saigon
Vietnamese Communist Party, 159
village council, 150, 151
vocabulary. *See* language
void decks, 94–95

Wakil (Bangladeshi businessman), 70

Walker, Brett, 6
walking city, 31, 33
Wang Hongxin, 188–89
Wang Li, 186n9
Warde, Paul, 11
wastelands, 191
waste management, 2, 7, 127–28, 168, 169
water: assetizing Gurgaon's, 142, 147–51; Gurgaon's lack of, 138–39; Saigon residents on city's, 165. *See also* flooding; monsoons
water bodies: defining, 143–47; disappearance of, 140–41, 152; drainage causes flooding in Saigon, 162; in Thủ Thiêm, 169, 170; inventory of, *145 table 7.1*, 153; lakes, 12, 13, 140, 187–88 (*See also* Ghata Jheel (Ghata Lake)); movement to restore, 152–53; traditional Southeast Asian housing along, 167. *See also* ponds; rivers; wetlands and marshes
water management, 70–71, 159, 164, 171
Water Management Department (Taipei), 128
water tanks, 12–13
Wei (Grange employee), 178
West Bengal. *See* climate change adaptation (Sundarban region)
Western worldviews, 3, 85n7, 88–89, 90. *See also* colonialism; Europe and European worldviews
West Lake, 187–88
West Lake District, 177, 183, 184, 185, 187
wetlands and marshes: in India, 7, 17, 149–50; paved over in Saigon, 160, *161 fig. 8.1*, 166, 172, 173; Xixi Wetland, 178, 187–90. *See also* water bodies
Whatmore, Sarah, 97
WhatsApp, 113
Whitington, Jerome, 164
wilderness spirits, 33–34
"wild" food, 91–92
Wildlife in the Anthropocene (Lorimer), 96–97
wildness, 182, 189–90
Williams, Raymond, 143

wind, 171
Winichakul, Thongchai, 34
witches, 31–32
women: as cat feeders, 96; bodily injuries and, 108; Chinese migrants to Singapore, 95
Woolfson, Esther, 14
workshops, 85–86
World Bank, 64, 159n1
World Wildlife Fund (WWF), 64–65, 75–76
worship, pothole, 110, *111 fig. 5.1*, 112
Wuya (Zhu Ziping), 95

Xiaobitan (community), 129–30
Xindian River, 124, 126, 128, 129, 130–32
Xixi Wetland, 178, 187–90

Yadavs (caste), 148, 151
Ya Nak (spirit), 29
Yeoh, Brenda S., 92
You Qing, 128
Yuen, Belinda, 83
Yunquan Village, 183

ZED group, 53
Zhou Xi-wei, 131
Zhu Ziping, 95